TISSUE ENGINEERING

Tissue Engineering

Current Perspectives

Tissue Engineering

Current Perspectives

Eugene Bell
Editor

Birkhäuser
Boston • Basel • Berlin

Eugene Bell
Marine Biological Laboratory
Woods Hole, MA 02543
and
Tissue Engineering Inc.
Cataumet, MA 02534

Library of Congress Cataloging-in-Publication Data

Tissue engineering : current perspectives / Eugene Bell, editor.
p. cm.
Includes bibliographical references and index.
ISBN 0-8176-3687-0 (acid-free paper : Boston).—ISBN 3-7643-3687-0 (acid-free paper : Berlin)
1. Animal cell biotechnology—Congresses. 2. Tissue culture—Congresses. I. Bell, Eugene
TP248.27.A46T58 1993
660'.6—dc20 93-28004
CIP

Printed on acid-free paper.

Birkhäuser 

ISBN 0-8176-3687-0
ISBN 3-7643-3687-0

Camera-ready copy prepared by the authors
Printed and bound by Quinn–Woodbine, Inc., Woodbine, NJ
Printed in the United States of America

9 8 7 6 5 4 3 2 1

Contents

V MATERIALS FOR TISSUE REMODELING *IN VIVO* AND *IN VITRO*

VI APPROACHES TO ALLOGRAFTING ENGINEERED CELLS AND TISSUES

Acknowledgements

I would like to express thanks to the Keystone Organization and to Fred Fox whose intuition that we collaborate with the Integrin and Cell Factors symposia by planning joint sessions resulted in an enriched Tissue Engineering Meeting held at Keystone, Colorado in 1992. I would like to thank Paul Mahotz and his assistant Denny Millard for their labors, their time and attention in helping with the organization of the meeting. The Tissue Engineering Symposium was co-sponsored by Genentech, Inc. and the Keystone Symposia on Molecular and Cellular Biology, through the Director's Sponsor fund established with gifts from Icos Corporation, Monsanto Corporation, Shering Corporation, and the Warner-Lambert Company. The National Science Foundation generously funded the proposal we submitted for support of the Tissue Engineering Symposium, as did the National Institute of Health. We owe sincere thanks to Dr. Fred Heiniken, program director of Biotechnology at the NSF, to Dr. Yvonne Maddox, Deputy Director of the Biophysics and Physiological Science Program of the NIGMS, NIH, and to Dr. Denuta Krotowski, Branch Chief for Basic Rehabilitation Medicine Research, NCRR, NIH. They have been ardent supporters of the infant but emerging discipline of tissue engineering at their respective institutions.

This is the third Tissue Engineering Symposium held under the aegis of the Keystone Organization. Dick Skalak, Fred Fox, and Randall Schwartz were organizers of the last meeting in 1990; Fred Fox, Dick Skalak, Fred Heiniken and Bert Fung were the prime movers for the first meeting held in 1988. They built a strong foundation for the growing TE edifice.

Royalties that flow from the publication of this volume have been donated to the Keystone Organization.

Eugene Bell

Contributors

R. Wayne Alexander, Division of Cardiology, Emory University School of Medicine, P.O. Box Drawer LL, Atlanta, GA 30322, USA

Dianne L. Atkins, Department of Pediatrics, Division of Pediatric Cardiology, University of Iowa, 200 Hawkins Drive, Iowa City, Iowa 52242-1083, USA

Stephen F. Badylak, Hillenbrand Biomedical Engineering Center, Purdue University, 1293 Potter Building, Room 204, West Lafayette, Indiana 47907-1293, USA

Eugene Bell, Tissue Engineering, Inc., 1248 Rt. 28A, Unit 6, Cataumet, MA 02534, USA

Bradford C. Berk, Division of Cardiology, Emory University School of Medicine, P.O. Box Drawer LL, Atlanta, GA 30322, USA

Helen M. Blau, Department of Pharmacology, Stanford University School of Medicine, 300 Pasteur Drive, Stanford, California 94305-5332, USA

Marie-José Blouin, Ludwig Institute for Cancer Research, University College and Middlesex Hospital, School of Medicine Branch, Courtauld Building, 91 Riding House St., London W1P 8BT, United Kingdom

Mary Bartlett Bunge, The Miami Project, University of Miami, 1600 NW 10th Avenue, R-48, Miami, Florida 33136, USA

Robert M. Carr, Organogenesis, Inc., 150 Dan Road, Canton, Massachusetts 02021, USA

John F. Cavallaro, Organogenesis, Inc., 150 Dan Road, Canton, Massachusetts 02021, USA

Claude Chamberland, Centre de Recherche, L'Hôtel-Dieu de Québec, 11 Côte du Palais, Québec City, Québec, Canada G1R 2J6

Elaine C. Davis, Department of Cell Biology and Physiology, Washington University School of Medicine, Campus Box 8228, 660 S. Euclid Avenue, St. Louis, Missouri 63110, USA

Jyotsna Dhawan, Department of Pharmacology, Stanford School of Medicine, 300 Pasteur Drive, Stanford, California 94305-5332, USA

Bruce A. Ekstein, Department of Respiratory Biology, Harvard School of Public Health, 665 Huntington Ave., Boston, Massachusetts 02115, USA

Christina Fernandez-Valle, The Miami Project, University of Miami, 1600 NW 10th Avenue, R-48, Miami, Florida 33136, USA

John A. Frangos, Department of Chemical Engineering, The Pennsylvania State University, 150 Fenske Building, University Park, Pennsylvania 16802, USA

Yuan-Chen Fung, Department of Bioengineering, 0412, University of California, San Diego, 9500 Gilman Drive, La Jolla, California 92093-0412, USA

Jose P. Garcia, Department of Surgery, West Roxbury VA Medical Center, 1400 VFW Parkway, West Roxbury, Massachusetts 02132, USA

Geoffrey Goldspink, Department of Anatomy and Developmental Biology, The Royal Free Hospital School of Medicine, Rowland Hill St., London University, NW3 2PF, United Kingdom

Farshid Guilak, Musculoskeletal Research Laboratory, Health Sciences Center T18-030, State University of New York, Stony Brook, Stony Brook, New York 11794, USA

Hsyue-Jen Hsieh, Department of Chemical Engineering, The Penn State University, 150 Fenske Building, University Park, Pennsylvania 16802, USA

Donald Ingber, Harvard Medical School, The Children's Hospital, Departments of Surgery and Pathology, Enders 1007, 300 Longwood Avenue, Boston, Massachusetts 02115, USA

Seigo Izumo, Molecular Medicine Division, Beth Israel Hospital, 330 Brookline Avenue, Boston, Massachusetts 02215, USA

Olena Jacenko, Anatomy and Cellular Biology, Harvard Medical School, 220 Longwood Avenue, Boston, Massachusetts 02115, USA

Paul D. Kemp, Organogenesis, Inc., 150 Dan Road, Canton, Massachusetts 02021, USA

Shukri F. Khuri, Department of Surgery, West Roxbury VA Medical Center, 1400 VFW Parkway, West Roxbury, Massachusetts 02132, USA

Nan-Qian Li, Department of Chemical Engineering, The Penn State University, 150 Fenske Building, University Park, Pennsylvania 16802, USA

Shu-Qian Liu, Department of Bioengineering, 0412, University of California, San Diego, 9500 Gilman Drive, La Jolla, California 92093-0412, USA

Anne Loranger, Centre de Recherche, L'Hôtel-Dieu de Québec, 11 Côte du Palais, Québec City, Québec, Canada G1R 2J6

Phyllis A. LuValle, Anatomy and Cellular Biology, Harvard University Medical School, 220 Longwood Avenue, Boston, Massachusetts 02115, USA

Normand Marceau, Centre de Recherche, L'Hôtel-Dieu de Québec, 11 Côte du Palais, Québec City, Québec, Canada G1R 2J6

Robert P. Mecham, Department of Cell Biology and Physiology, Washington University School of Medicine, Campus Box 8228, 660 S. Euclid Avenue, St. Louis, Missouri 63110, USA

Richard G. Miller, Cell and Molecular Biology, Ontario Cancer Institute, 500 Sherbourne Street, Toronto, Ontario, Canada M4X 1K9

Masako Mitsumata, School of Mechanical Engineering, Georgia Institute of Technology, Corner of Cherry and 1st Street, SST Building, Room 300, Atlanta, Georgia 30332-0405, USA

Van C. Mow, Orthopedic Research Laboratory, Columbia University, 630 West 168th Street, Black Building, Room 1412, New York, New York 10032, USA

Robert M. Nerem, School of Mechanical Engineering, Georgia Institute of Technology, Corner of Cherry and 1st Street, SST Building, Room 300, Atlanta, Georgia 30332-0405, USA

Micheline Noël, Centre de Recherche, L'Hôtel-Dieu de Québec, 11 Côte du Palais, Québec City, Québec, Canada G1R 2J6

Kimberlie D. O'Neil, Organogenesis, Inc., 150 Dan Road, Canton, Massachusetts 02021, USA

Bjorn R. Olsen, Anatomy and Cellular Biology, Harvard Medical School, 220 Longwood Avenue, Boston, Massachusetts 02115, USA

Carlos L. Paino, Servicio de Bioquímica-Investigación, Hospital Ramón y Cajal, Carretera de Clomenar, km. 9, 28034 Madrid, Spain

Grace K. Pavlath, Department of Pharmacology, Stanford School of Medicine, 300 Pasteur Drive, Stanford, California 94305-5332, USA

Mireille Rosenberg, Genzyme Corporation, One Kendall Square, Cambridge, Massachusetts 02139, USA

Jun-ichi Sadoshima, Molecular Medicine Division, Beth Israel Hospital, 330 Brookline Avenue, Boston, Massachusetts 02115, USA

Richard Skalak, Institute of Biomedical Engineering, University of California, San Diego, 9500 Gilman Drive, La Jolla, California 92093-0412, USA

Michael Tantillo, Department of Surgery, West Roxbury VA Medical Center, 1400 VFW Parkway, West Roxbury, Massachusetts 02132, USA

Ronald G. Tompkins, Massachusetts General Hospital, Bigelow 1302, Boston, Massachusetts 02114, USA

Mehmet Toner, Surgical Services, Massachusetts General Hospital, CNY-5, Charlestown, Massachusetts 02114, USA

Dan W. Urry, Laboratory of Molecular Biophysics, The University of Alabama at Birmingham, 525 Volker Hall, VH 300, UAB Station, Birmingham, Alabama 35294-0019, USA

Crispin B. Weinberg, 25 Beals Street, Brookline, Massachusetts 02146, USA

Martin Yarmush, Massachusetts General Hospital, Bigelow 1302, Boston, Massachusetts 02114, USA

Thierry Ziegler, School of Mechanical Engineering, Georgia Institute of Technology, Corner of Cherry and 1st Street, SST Building, Room 300, Atlanta, Georgia 30332-0405, USA

Section I

Overview of Tissue Engineering

TISSUE ENGINEERING, AN OVERVIEW

Eugene Bell
Marine Biological Laboratory and CEO and Chief Scientist, Tissue Engineering Inc., Woods Hole, Massachusetts, 02543

In organizing the present Symposium, I have leaned mainly on my own conception of what Tissue Engineering embraces. There are surely more views than one that will be tempered and melded as the field grows. Tissue engineering can be thought of in terms of its goals, namely the following: 1) providing cellular prostheses or replacement parts for the human body; 2) providing formed acellular replacement parts capable of inducing regeneration; 3) providing tissue or organ-like model systems populated with cells for basic research and for many applied uses such as the study of disease states using aberrant cells; 4) providing vehicles for delivering engineered cells to the organism; and 5) surfacing non-biological devices. Discussion of each of the foregoing goals and examples of work toward them make up Part I of this overview. Citations from the experience of others and our own are by no means inclusive. Part II deals with additional examples; primarily it deals with specific areas of work that are or may become underpinnings of this new field.

I. GOALS

Providing cellular replacement parts

In the '80s, we developed a Living Skin Equivalent (LSE) that consisted of a dermal component with a fully developed epidermis (Bell et al. 1979, 1983, 1991). The dermal equivalent (DE) was constituted by mixing dermal fibroblasts in a solution containing serum, culture medium, and collagen at a concentration of 1mg/ml. At neutral pH and with warming to 37° a gel or lattice formed in which the cells were uniformly distributed. The cells interacted with the collagen fibrils condensing them and in the process causing fluid to be expressed from the lattice. The LSE was successfully implanted as a skin substitute in experimental animals (rats) remaining functional for the lifetime of the hosts (Bell et al. 1981, 1984b,c; Hull et al. 1983). The LSE could be pigmented by addition of melanocytes (Topol et al. 1986). Melanocytes were plated on the DE after it had contracted and before keratinocytes were added. The melanocytes became functionally active, donating pigment to the keratinocytes that covered them.

The DE and the dermal-epidermal junction of the LSE were both enriched by biosynthetic contributions from dermal fibroblasts (Nusgens et al. 1984) and keratinocytes, respectively, although basal lamina formation (Topol et al. 1986) was not continuous. The biosynthetic activity of the cells in the DE could be substantially stimulated by additions (e.g. ascorbate and growth factors) to the culture medium (Bell et al. 1991). Similarly, and also in vitro, the LSE developed a fully differentiated epidermis in which the known markers of phenotypic specificity of the basal, spinous, and granular layers of the skin were expressed (Parenteau et al. 1991). The LSE also developed a functional stratum corneum with barrier properties within an order of magnitude of those of actual skin. A version of the LSE for testing, TESTSKIN, was commercialized by Organogenesis Inc., the company I founded and retired from; another version of the LSE for grafting is in clinical trials. A significant shortcoming of the LSE for grafting was the lack of strength and oversimplification of the DE and the incompletely developed dermal-epidermal junction.

In addition to the LSE, my colleagues and I developed a living artery equivalent (Weinberg and Bell 1986), a thyroid gland equivalent (Bell et al. 1984a), adipose tissue (Kagan and Bell 1989), endothelium from the CNS with barrier properties (Bell et al. 1990) and other tissues including corneal, cardiac tissue, and intestinal epithelium (unpublished results). We found that in vitro it is possible to make instant "mesenchyme" with cultivated and banked connective tissue cells like fibroblasts or adipocytes or keratocytes of the cornea or smooth muscle cells by combining the cells with a collagen-containing cocktail which becomes a polymeric network or lattice that, once formed, the cells can condense and enrich. The formed "mesenchymal" tissue, as a substratum, can be over or underplated with endothelium, mesothelium, or epithelium; or in addition to the contractile connective tissue cells used to condense it, it can also be seeded at the outset with cells that form tubules, acini, or follicles. This is the approach we took in constituting a thyroid gland equivalent with long term cultured thyroid cells (Bell et al. 1984a) and managed to rescue thyroidectomized rat hosts who received the grafts. What was remarkable was the differentiation of thyroglobulin filled follicles in the ectopically placed glandular equivalents. An important principal we discovered is that limited differentiation of an engineered living tissue construct in vitro may not be an indicator of its in vivo potential, particularly in an organism expressing a deficiency. In the thyroidectomized animal, the high level of TSH was the humeral stimulus needed to induce differentiation in the implant, while in normal animals no development of implants was observed.

The foregoing tissue-like constructs were populated with one or more cell types typical of the tissue in vivo. This approach, using cells of the mammary gland on collagen substrata, has been pioneered by Bissell and coworkers (e.g. Streuli and Bissell 1990). The subjects of several of the contributions in this volume fall into the category of cellular replacement parts (Bunge et al., Chapter 5, this volume; Toner et al., Chapter 9, this volume). Materials other than animal biopolymers such as collagen, but also being bioabsorbable like poly L-lactic acid, polyglycolic acid and combinations of them, have been seeded with cells in vitro and used as transplants (Mooney et al. 1992). While the preceding examples deal with replacement parts reconstituted in vitro, primarily with cultured cells and matrix components, there is a long history of use by reconstructive surgeons of body parts removed from one location to serve a new function in another (Goldspink, Chapter 16, this volume).

Providing formed acellular replacement parts capable of inducing regeneration

Providing formed materials capable of inducing regeneration of tissues and of being remodeled or absorbed is not a new approach in tissue engineering. Finding that demineralized bone can induce regeneration of cartilage and bone tissues ectopically (Reddi and Huggins 1972; Urist 1965) led eventually to the discovery of several active factors including a recombinant human bone morphogenetic protein (rhBMP-2) capable of healing segmental defects in dogs and sheep when combined with inactive demineralized bone matrix (Wang et al. 1992). Substances free of growth and differentiation hormones—a cementitious calcium phosphate material for example—also have the capacity to induce repair and be completely remodeled (Constantz et al. 1992). Vascular and soft tissue orthopedic prostheses constituted from the small intestinal submucosa of the pig have been implanted xenogeneically in dogs and monkeys (Badylak, Chapter *17, this volume). Work I directed at the company Organogenesis, based on the use of dense fibrillar collagen tubes, continued after my departure. The tubes, designed and tested in animals as vascular grafts, are reported to undergo extensive remodeling (Weinberg et al., Chapter 18, this volume). Earlier work of van der Lei and colleagues (1987) using a co-polymer biodegradable small caliber tube in rats reported remodeling of the biomaterial and remarkable induction of arterial regeneration.

Providing tissue or organ-like model systems populated with cells for basic research and for many applied uses such as toxicity testing with normal cells and the study of disease states using aberrant cells

TESTSKIN, referred to above, has been used widely for testing many substances that come into contact with the skin (Bell et al. 1989a, 1989b). The LSE model as an organotypic system has also been used to study a disease state using aberrant cells. For example, with a group in Paris (Saiag et al. 1985), we developed a psoriatic skin model with which it was demonstrated that psoriatic fibroblasts in a dermis-like collagen lattice were able to induce a principal feature of the disease, hyperproliferation, in normal keratinocytes plated over the dermis to form a differentiating epidermis.

Models of a three layered artery (Weinberg and Bell 1986), made up with three cell types, endothelium, smooth muscle and advential fibroblasts, organized into intima, media and adventitia may be especially useful for studying the induction of atherosclerosis. Recently, the LSE was used to develop a wound healing model (Bell and Scott 1992). Discs of skin equivalent were punched out of LSEs and the cylindrical dermal equivalent gaps were filled with various cell-matrix or matrix components and the rate of overgrowth of the filled wound by peripheral epidermis was studied, wound closure being the end point of the assay. The psoriasis, atherosclerosis and wound healing models referred to can serve a second purpose, as systems useful for devising therapies for the treatment of the diseases or injury in question.

Three dimensional model systems with or without supporting cells have a place as diagnostic devices in cancer chemotherapy to select kinds and regimens of use of chemotherapeutic agents, especially if they can promote growth of all types of transformed cells present in tumor biopsies. Since three dimensional model systems populated with cells form tissues and differentiate in vitro, they can be used for the cultivation of viruses. Papilloma virus might be cultivated in a system such as the Living Skin Equivalent.

Providing vehicles for delivering engineered cells to the organism

Making the right kind of proliferating cell dopaminergic by inserting the tyrosine hydroxylase gene, or finding the neuroblast that can differentiate along the appropriate phenotypic pathway to become dopaminergic, would be steps on the road to treating Parkinson's disease. But delivering cells to the CNS in a manner that will localize them and insure their persistence and function may require a transient matrix that can be rapidly vascularized and programmed to disappear.

Surfacing non-biological devices

Finally, one of the goals of tissue engineering is the surfacing of non-biological devices such as biosensors, smart devices, catheters, and orthopedic prostheses with living tissues or materials that can give cells that are mobilized from the surrounds a chance to win the race for the surface that is normally won by microorganisms. Designing accommodating tissue interfaces applied to non-biological devices before implantation may improve their functionality and reduce complications often due to infection, but also due to the mismatch of living tissues and the prosthesis or device to which they are apposed.

II. SPECIFIC AREAS OF WORK

To be able to reconstitute and successfully implant tissues in humans is a formidable challenge. Recognizing the goals of tissue engineering directs our attention to several specific areas of work on whose progress tissue engineering depends. They include the following: 1) establishing reserves of cells needed for implants designed to restore a lost function, because the cells which normally perform the function are gone or damaged; 2) understanding the composition and design of extracellular matrices, their ligands and the cell receptors, the integrins, that recognize them; 3) identifying factors which play a role in the development of tissues, namely chemical agents such as growth hormones and physical forces; discovering ways of applying physical forces in vitro to recapitulate the events of histogenesis; 4) learning how to implant immunologically foreign tissues across major histocompatibility barriers; and 5) defining criteria for selecting and testing matrices best suited for specific replacement prostheses. Each of the foregoing is discussed below.

Cells

Some specialized normal human cells can be cultivated readily and can be scaled up for banking. A partial list includes: skin cells, vascular cells, adipose tissue cells, skeletal muscle cells, chondrocytes, osteoblasts, mucogingeval cells, corneal cells, skeletal muscle cells (satellite cells), and pigment cells. The cultivation of other cell types still represents a challenge, but there is growing confidence that the body harbors a variety of stem cells now being identified by new techniques (Friedman and Weissman 1991; Blau et al., Chapter 4, this volume; Githens 1993; Hall and Watt 1989; Lendahl et al. 1990; Marceau et al., Chapter 6, this volume; McKeehan et al. 1990; Renfranz et al. 1991; Uchida and Weissman 1992; Wren et al. 1992; Goldspink, Chapter 16, this volume). A recent comment on the growing belief in the reality of the liver stem cell (Travis 1992) is relevant. Isolating human hematopoietic stem cells, neuroblasts, exocrine glandular cells, such as liver and salivary gland, cells of the intestinal mucosa, endocrine glandular cells such as pancreatic islet cells and parathyroid cells and others are high priority needs for tissue engineering. The propagation of stem cells will have a profound impact on the treatment of many disorders. New growth cocktails, new substrates, cocultures and scale-up methodologies have in themselves led to successful propagation of difficult to grow human cells such as keratinocytes, neuroblasts, striated muscle cells, thyroid and parotid gland cells and will continue to be important in providing needed cells for Tissue Engineering.

Matrix

The composition and properties of the extracellular matrix (ECM) have become subjects of broad interest. Extracellular matrix and the three dimensional structure of tissues are crucial determinants in the chain of events leading to genetic readout on which differentiation depends (Bissell and Barcellos-Hoff 1987). Cues from the ECM can be thought of as initiators of intracellular signaling and chemical reactions governing the specificity of cell differentiation and function (Ingber, Chapter 7, this volume). It has become clear that the mechanical properties of tissues and organs depend largely on three dimensional collagen fiber frameworks that are tissue specific. The development and functioning of the frameworks depend in part on a newly discovered class of molecules called FACIT (Fiber Associated Collagen with Interrupted Triple helix) (Olsen et al., Chapter 2, this volume). Non-fibrillar collagen types IX, XII, and XIV belong to FACIT and are thought to provide binding sites for other matrix components or cells along the surface of collagen fibrils. The FACIT matrix ligands, the nectin ligands and the attachment sites of other matrix molecules serve as mooring sites for cells through the receptor system provided by trans-membrane proteins, the integrins (Albelda and Buck 1990; Hynes 1987; Pytela et al. 1985) and syndecan (Bernfield and Sanderson 1990). Interwoven through the ECM are the proteoglycans that even as minor components in arteries for example, exert a major influence on arterial viscoelasticity, permeability, lipid metabolism and thrombosis (Wight 1989). The complexity and composition of the proteoglycan content depends on whether the generating cells are quiescent or activated through exposure to regulatory factors. Both TGFb and PDGF increase biosynthesis of the proteoglycan versican, but only PDGF enhances proliferation of smooth muscle cells in the media of an artery. Non-sprouting endothelial cells don't make PG11/decorin or collagen type I, but sprouting cells do (Wight 1989). There is also a reciprocal regulation of growth factor action by matrix, since, for example, TGFb is inactivated by the proteoglycan decorin (Border et al. 1990; Nakamura et al. 1990; Okuda et al. 1990). In reviewing the community of the extracellular matrix, there is also a major class of molecules that make up the microfibrils important for the cross-linked network of elastin. A component microfibrillar element, fibrillin, contains RGD sequences to which cells can bind. Cells do bind to elastin; arterial endothelium is in contact with elastin before a basal lamina forms (Davis and Mecham, Chapter 3, this volume).

The ECM components include in addition to the structural elements a large number of growth and differentiation factors, cytokines, and other regulatory and signaling molecules found in association with the structural matrix. Given the breadth of causal influences affecting cell function and differentiation exerted by the ECM, one approach to tissue engineering, whether deliberate or not, has embraced the use of one or more ECM components for reconstituting a tissue in combination with cells, or for use alone as a prosthesis. Without added cells the expectation is that the prosthesis will induce regeneration by mobilizing cells from contiguous tissues and the circulation after being implanted (Badylak, Chapter 17, this volume; van der Lei et al. 1987; Weinberg, Chapter 18, this volume). The use of frozen or freeze dried tissues or minimally processed, that is not cross linked tissues has enjoyed important successes (Barad et al. 1982; Arnoczky et al. 1986; Gonzales-lavin et al. 1990; O'Brien et al. 1987). Augmentation of matrices of various kinds by the addition of ECM ligands (RGD) (Glass et al. 1992) or with growth and differentiation factors also have promise. Another approach, not new, makes use of man-made polymers that have no relation to the composition of the ECM. This category includes non-degradable prostheses, (Jarrell and Williams 1990; Nyman et al. 1987) using man-made synthetics, as well as degradable materials (Constantz et al. 1992; Mooney et al. 1992; van der Lei et al. 1987). Remodelable elastomeric polymers in the form of prostheses (Urry, Chapter 19, this volume) with built in cell attachment sites could provide transient scaffolds for mobilized host cells.

A number of ways have been used to prepare matrices and prostheses for tissue engineering: 1) monolayered cells can make and secrete matrix in vitro, matrigel for example (Kleinman et al. 1986); 2) matrix can be prepared by extracting and purifying collagens, and allowing cells mixed with the collagen before it gels, to condense it after gelation and to enrich it with secreted products, a process that can be enhanced by addition of growth and differentiation factors and by paracrine factors provided by co-cultured cells (Bell et al. 1991); 3) fibers or sheet material can be created from animal polymers, from artificial biodegradable biopolymers (Urry, Chapter 19, this volume; Vacanti et al. 1992), and from man-made non-biological polymers such as dacron, teflon and other synthetics to produce fabrics in various forms; 4) by fabricating tubes or sheets from any of the foregoing materials without using conventional textile approaches, but by using processes of condensation or molding (Weinberg et al., Chapter 18, this volume); 5) by synthesizing human structural and other molecules such as collagen (collagen analogs e.g.) (Goldberg et al. 1989), elastin or the nectins for example in microbial and other cell systems using recombinant DNA methodology. The approaches leave room for inserting regulatory and other factors in prostheses constituted with designer macromolecules. The usefulness of recombinant collagen would depend on the success with which post transcriptional processing, required for polymerization, can be achieved.

Looking into the future, imitating specific histogenesis in vitro may become easier than it is now—if the cells combined with the matrix can be put into the right state of differentiation. For sequences of cell-matrix and cell-cell interactions that must be played out. In vivo, they are preceded and accompanied by rounds of cell divisions and matrix biosynthesis and secretion. Feedback from the matrix stimulates the producer cells to further biosynthetic activity, but it can also stimulate cells to move, to form acini, tubules, follicles and other structures; to coalesce into foci, for example, that can interact with overlying cells as in hair or feather formation. The products that make their way into a matrix emanating from one subpopulation of cells may act on neighboring contiguous cells which in turn make their contribution to the organogenesis cascade. To drive development in vitro to a desired culmination what may be needed is the right constellation of cells, in cocultures, in order to promote tissue growth and/or differention (Atkins, Chapter 8, this volume). We found that a predesigned matrix with one type of cell, the dermal fibroblast, and a sequence of different medium compositions, each tailored to the changing needs of the of a second type of cell, the keratinocyte, were required to promote faithful organotypic development of the LSE (Parenteau et al. 1991).

The role of physical forces

To drive development in vitro, the design and engineering of tissue forming molds in which cell matrix combinations can be cast and tethered or simply tethered in order to be subjected to the shear, tensile and other forces known to occur in the course of development, may also be required. Physical forces acting at the right time in development may be as important epigenetic requirements for shaping tissues and initiating gene action as the autocrine, paracrine and endocrine factors with which cells supply themselves and one another. I cite as an example work of Ilizarov from the Kurgan All Union Center for Restorative Traumatology and Orthopaedics who has been elongating bones forcefully for forty years (reviews by Ilizarov, 1988, 1989, 1990; also see Goldspink, Chapter 16, this volume). More recently, he has been studying the elongation of soft tissues that occurs in the course of bone elongation brought about by the application of tension to the bone through an external armature fixed at two levels of the bone. Frequent extensions of the armature to yield a rate of extension of 1.0 mm/day resulted in the best outcome for perosseous tissues including skeletal muscle, blood vessels, connective tissue, nerve and skin all of which appear to be stimulated by the persistent application of tensile stress.

The role of mechanical stresses in the remodeling of cartilage (Mow and Guilak, Chapter 13, this volume), the control of growth of vascular endothelium (Nerem et al., Chapter 12, this volume), their effect on blood vessel walls (Fung and Liu, Chapter 11, this volume) point to them as important regulators of bio-processes. Gene expression induced by stress (Hsieh et al., Chapter 15, this volume) and the signaling pathways leading to gene activation and expression initiated by mechanical stretch, for example, are beginning to be understood (Izumo and Sadoshima, Chapter 14, this volume). An overview of the subject (Skalak, Chapter 10, this volume) suggests that growth, morphogenesis and differentiation of all tissues may be dependent on mechanical stress. It is well known of course that various forms of exercise including pumping iron can change the size shape and density of tissues, promoting growth and differentiation. Hormones and diet can also bring about significant tissue changes. The foregoing influences can be thought of as minimally invasive tissue engineering that can have an important future.

Immunologic responses to implanted engineered tissues

The success of delivering packaged cultivated cells to a recipient to restore a functional capacity lost through disease, accident, aging, or war depends in part on the way the implant is greeted by the host. The implant must be immunologically acceptable—in other words, neutral. Strategies for inducing host unresponsiveness to allogeneic grafts are abroad, and have been notably successful in experimental animal models (Heeg and Wagner 1990). Using reconstituted tissues and organs made up with selected cultivated allogeneic cells has provided a unique opportunity to dissect the immune target. It is some years now that we transplanted allogeneic skin cells in a living skin equivalent in which cells of the immune system and of the microcirculation were absent because we excluded them deliberately. Grafts made to young animals across major histocompatibility barriers persisted for the life span of the recipients as determined by karyotyping since grafts were constituted with female cells and were made to males (Sher et al. 1983). The idea that allograft rejection is initiated by passenger cells—that is, immune system cells present in most tissues (Bell and Rosenberg 1990; Billingham 1971; Lacy et al. 1979a, 1979b; Lafferty and Woolnough 1977; Rosenberg, Chapter 21, this volume; Sher et al. 1983; Steinmuller and Hart 1971) and how their elimination or neutralization can lead to graft acceptance—has gained wide support. New strategies for inducing T cell anergy or host unresponsiveness are among the growing possibilities for using cultivated allogeneic cells for transplantation without the need of immunosuppression (Fink et al. 1986; Heeg and Wagner 1990; Schwartz 1990; Miller, Chapter 20, this volume).

Finally keeping rigorous track of cells in engineered implants to account for cell number, cell persistence, integration and functionality is a real concern. In animal experiments, accountability by cell marking has been used successfully. Whether the installed cells are muscle cells now being used to treat muscular dystrophy (Blau et al., Chapter 4, this volume), or neuroblasts to determine their adaptability when implanted at various sites in the CNS (Renfranz et al. 1991), or dopaminergic cells to treat Parkinson's disease, or pancreatic islet cells used to treat diabetes, the arithmetic of immune acceptance, of persistence of cell number as a function of the number of cells grafted and of the number functioning in the graft with time, would be valuable to have. In general, the fate—that is, the failure or success of engineered implants in trial systems—is an issue very early in the evolution of new products. This applies not only to cells but to the materials with which they are associated as well. The potential of biomaterials, destined for tissue engineering use, to induce regeneration, to be remodeled, to be immunologically acceptable, and to develop the functional capacity characteristic of the tissue or organ being replaced are crucial endpoints. Choosing the right trial systems for testing them with cells in vitro and in vivo is crucial and would benefit from some degree of consensus and standardization.

References

Albelda SM, Buck CA (1990): Integrins and other cell adhesion molecules. *FASEB J* 4:2868

Arnoczky SP, Dipl DVM, Warren RF, Ashlock MA (1986): Replacement of the anterior cruciate ligament using a patellar tendon allograft. *Bone Joint Surg* 68-A:376

Barad S, Cabaud HE, Rodrigo JJ (1982): The effect of storage at -80°C as compared to 4°C on the strength of Rhesus monkey anterior cruciate ligament. *Trans Orthop Res Soc* 7:378

Bell E, Ehrlich HP, Buttle DJ, Nakatsuji T (1981): Living tissue formed in vitro and accepted as skin equivalent tissue of full thickness. *Science* 211:1052

Bell E, Gay R, Swiderek M, Class T, Kemp R, Green G, Haimes H, Bilbo P (1989a): Use of fabricated living tissue and organ equivalents as defined higher order systems for the study of pharmacologic responses to test substances. In: *Pharmaceutical Application of Cell and Tissue Culture*, Davis SS, Illum L, eds. London: Plenum Publishing Co., Ltd.

Bell E, Ivarsson B, Merrill C (1979): Production of a tissue-like structure by contraction of collagen lattices by human fibroblasts of different proliferative potential in vitro. *Proc Natl Acad Sci USA* 76:1274

Bell E, Kagan D, Nolte C, Mason V, Hastings C (1990): Demonstration of barrier properties exhibited by brain capillary entothelium cultured as a monolayer on a tissue equivalent. *J Cell Biol* 5:185a

Bell E, Moore H, Michie C, Sher S, Hull B, Coon H (1984a): Reconstitution of a thyroid gland equivalent from cells and matrix materials. *J Exptl Zool* 232:277

Bell E, Rosenberg M (1990): The commercial use of cultivated human cells. *Transplantation Proc* 22:971

Bell E, Rosenberg M, Kemp P, Gay R, Green G, Muthukumaran N, Nolte C (1991): Recipes for reconstituting skin. *J Biomech Eng* 113:113

Bell E, Rosenberg M, Kemp P, Parenteau N, Haimes H, Chen J, Swiderek M, Kaplan F, Kagan D, Mason V, Boucher L (1989b): Reconstitution of living organ equivalents from specialized cells and matrix biomolecules. In: *Organs Artificiels Hybrides*, Baquay C, Dupuy B, eds. Colloque Inserm 177:13

Bell E, Scott S (1992): Tissue fabrication: Reconstitution and remodeling in vitro. *Mat Res Soc Symp Proc* 252:141

Bell E, Sher S, Hull B (1984b): The living skin equivalent as a structural and immunological model in skin grafting. *Scanning Electron Microscopy* 4:1957

Bell E, Sher S, Hull B, Merrill C, Rosen S, Chamson A, Asselineau D, Dubertret L, Coloumb B, Lapiere C, Nusgens B, Neveux Y (1983): The Reconstitution of living skin. *Journal of Invest Derm* 81:2s

Bell E, Sher S, Hull B, Sarber R, Rosen S (1984c): Long term persistence in experimental animals of components of skin-equivalent grafts fabricated in the laboratory. In: *Eukaryotic Cell Cultures, Basics and Applications*, Acton RT, Lynn JD, eds. New York: Plenum Press

Bernfield M, Sanderson RD (1990): Syndecan, a developmentally regulated cell surface proteoglycan that binds extracellular matrix and growth factors. *Phil Trans R Soc Lond Ser B* 327:171

Billingham RE (1971): The passenger cell concept in transplantation immunology. *Cell Immunol* 2:1

Bissell MJ, Barcellos-Hoff MH (1987): The influence of extracellular matrix on gene expression: Is structure the message? *J Cell Sci Suppl* 8:327

Border W, Okuda S, Languino LR, Sporn MB, Rouslahti E (1990): Suppression of experimental glomerulonephritis by antiserum against TGF-β. *Nature* 346:371

Constantz BR, Young SW, Kienapfel H, Dahlen BL, Summer DR, Turner TM, Urban RM, Galente JO, Goodman SB, Gunasekaran S (1992): Calcium phosphate cement in a rabbit femoral canal model and a canine humeral plug model: A pilot investigation. *Mat Res Symp Proc* 252:79

Fink PJ, Shimenkovitz RP, Bevin MJ (1986): Veto cells. *Ann Rev Immunol* 6:115

Friedman J, Weissman IL (1991): Mouse hematopoietic stem cells. *Blood* 78:1395

Githens SJ (1993): Pancreatic development in animals. In: *The Pancreas, Biology, Pathobiology, and Disease, 2nd ed.*, Go VLW, ed. New York: Raven Press

Glass JR, Craig WS, Dickerson K, Pierschbacher MD (1992): Cellular interaction with biomaterials modified byh arg-gly-asp containing peptides. *Mat Res Soc Symp Proc* 252:331

Goldberg I, Salerno AJ, Patterson T, Williams JI (1989): Cloning and expression of a collagen-analog-encoding synthetic gene in Escherichia coli. *Gene* 80:305

Gonzalez-lavin L, Spotnitz AJ, Mackenzie JW, Gu J, Gadi IK, Gullo J, Boyd C, Graf D (1990): Homograft valve durability: Host or donor influence. *Heart Vessels* 5:102

Hall PA, Watt FM (1989): Stem cells: The generation and maintenance of cellular diversity. *Development* 106:619

Heeg K, Wagner H (1990): Induction of peripheral tolerence to class I major histocompatibility complex (MHC) alloantigens in adult mice: Transfused class I MHCV-incompatible splenocytes veto clonal responses of antigen-reactive Lyt-2+ T cells. *J Exp Med* 172:719

Hull B, Sher S, Rosen S, Church D, Bell E (1983): Fibroblasts in isogeneic skin equivalents persist for long periods after grafting. *J Invest Derm* 81:436

Hynes R (1987): Integrins: A family of cell surface receptors. *Cell* 48:549

Ilizarov GA (1988): The principles of the Ilizarov method. *Bull Hosp Jt Dis Orthop Inst* 48:1

Ilizarov GA (1989): The tension-stress effect on the genesis and growth of tissues: Part II. The influence of the rate and frequency of distraction. *Clin Orthop* 239:263

Ilizarov GA (1990): Clinical application of the tension-stress effect for limb lengthening. *Clin Orthop* 250:8

Jarrell BE, Williams SK (1990): Creation of man-made endothelial cell linings. *J Cell Biochem* 14E:237

Kagan D, Bell E (1989): Differentiation of a fatty connective tissue equivalent without chemical induction. *J Cell Biol* 107:603a

Kleinman H, McGarvey ML, Hassel JR, Star VL, Cannon FB, Laurie GW, Martin GR (1986): Basement membrane complexes with biological activity. *Biochemistry* 25:312

Lacy PE, Davie JM, Finke EH (1979a): Prolongation of islet allograft survival following in vitro culture. *Science* 204:323

Lacy PE, Davie JM, Finke EH (1979b): Induction of rejection of successful allografts of rat islets by donor peritoneal exudate cells. *Transplantation* 28:415

Lafferty KJ, Cooley MA, Woolnough J, Walker KZ (1975): Thyroid allograft immunogenmicity is reduced after a period in organ culture. *Science* 188:259

Lafferty KJ, Woolnough J (1977): The origin and mechanism of the allograft reaction. *Immunol Rev* 35:231

Lendahl U, Zimmerman L, McKay RDG (1990): CNS stem cells express a new class of intermediate filament protein. *Cell* 60:585

McKeehan WL, Barnes D, Reid L, Stanbridge E, Murakami H, Sato GH (1990): Frontiers in mammalian cell culture. *In Vitro cell dev biol*, 26:9

Mooney JD, Cima L, Langer R, Johnson L, Hansen LK, Ingber DE, Vacanti JP (1992): Induction of hepatocyte differentiation by extracellular matrix and an RGD-containing synthetic peptide. *Mat Res Soc Symp Proc* 252:345

Nakamura T, Okuda S, Miller D, Rouslahti E, Border W (1990): Transforming growth factor-β (TGF-β) regulates production of extra-cellular matrix (ECM) components by glomerular epithelial cells. *Kidney Int* 37:221

Nusgens B, Merrill C, Lapiere C, Bell E (1984): Collagen biosynthesis by cells in a tissue equivalent matrix in vitro. *Collagen & Related Res* 4:351

Nyman S, Gottlow J, Lindhe J, Karring T, Wennstrom J (1987): New attachment formation by guided tissue regeneration. *J Periodont Res* 22:252

O'Brien M, Strafford E, Gardner M, Pohlner PG, McGiffin DC (1987): A comparison of aortic valve replacement with viable cryopreserved and fresh allograft valves with a note on chromosomal studies. *Thorac Cardiovasc Surg* 94:812

Okuda S, Languino LR, Rouslahti E, Border W (1990): Elevated expression of TGF-β and proteoglycan production in experimental glomerulonephritis. Possible role in expansion of mesangial ECM. *J Clin Invest* 86:453

Parenteau N, Nolte C, Bilbo P, Rosenberg M, Wilkins L, Johnson E, Watson S, Mason V, Bell E (1991): Epidermus generated *in vitro*: Practical considerations and applications. *J Cell Biochem* 45:245

Pytela R, Pierschbacher MD, Rouslahti E (1985): A 125/115 kDa cell surface receptor specific for vitronectin interacts with the arginine-glycine-aspartic acid adhesion sequence derived from fibronectin. *Proc Natl Acad Sci USA* 82:5766

Reddi H, Huggins C (1972): Biochemical sequences in the transformation of normal fibroblasts in adolescent rat. *Proc Natl Acad Sci USA* 69:1601
Renfranz PJ, Cunningham MG, McKay RDG (1991): Region specific differentation of the hippocampal stem cell line HiB5 upon implantation into the developing mammalian brain. *Cell* 66:1
Saiag P, Coulomb B, Lebreton C, Bell E, Dubertret L (1985): Psoriatic fibroblasts induce hyperproliferation of normal keratinocytes in a skin equivalent model in vitro. *Science* 230:669
Schwartz RH (1990): A cell culture model for T lymphocyte clonal anergy. *Science* 248:1349
Sher S, Hull B, Rosen S, Church D, Friedman L, Bell E (1983): Acceptance of allogeneic fibroblasts in skin equivalent transplants. *Transplant* 36:552
Steinmuller D, Hart EA (1971): Passenger leucocytes and induction of allograft immunity. *Transplant Proc* 3:673
Streuli CH, Bissell MJ (1990): Expression of extracellular matrix components is regulated by substratum. *J Cell Biol* 110:1405
Topol B, Haimes H, Dubertret L, Bell E (1986): Transfer of melanosomes in a skin equivalent model in vitro. *J Invest Derm* 87:642
Travis J (1992): The search for liver stem cells picks up. *Science* 259:1829
Uchida N and Weissman IL (1992): Searching for hematopoietic stem cells: Evidence that Thy-1.1^{lo} Lin$^-$ Sca-1$^+$ cells are the only stem cells in C57BL/Ka$^-$ Thy-1.1 bone marrow. *J Exp Med* 175:175
Urist MR (1965): Bone: Formation by autoinduction. *Science* 150:893
Vacanti JP, Cima LG, Radkowski D, Upton J, Vacanti JP (1992): Tissue engineered growth of new cartilage in the shape of a human ear using synthetic polymers seeded with chondrocytes. *Mat Res Symp Proc* 252:367
van der Lei B, Wildevuur RH, Nieuwenhuis P (1987): Long term biologic fate of neoarteries regenerated in microporous, compliant, biodegradable, small caliber vascular grafts in rats. *Surgery* 101:459
Wang EA, Toriumi DM, Gerhart TN (1992): Healing large segmental defects in dogs and sheep with recombinant human bone morphogenetic protein (rhBMP-2). *Materials Research Society Symposium Series* 252:267
Weinberg C, Bell E (1986): A blood vessel model constructed from collagen and vascular cells. *Science* 231:397
Wight TN (1989): Cell Biology of arterial proteoglycans. *Arteriosclerosis* 9:1
Wren D, Wolssijk G, Noble MJ (1992): In vitro analysis of the origin and maintenance of O-2A^{adult} progenitor cells. *J Cell Biol* 116:167

Section II

Matrix Molecules and Their Ligands

ROLE OF NON-FIBRILLAR COLLAGENS IN MATRIX ASSEMBLIES

Bjorn Reino Olsen, Phyllis A. LuValle, and Olena Jacenko
Department of Anatomy and Cellular Biology
Harvard Medical School

Extracellular matrices are composed of macromolecular components that interact to form highly complex supramolecular structures. Among these components are collagen fibrils. The fibrils represent tissue elements of high tensile strength, and their tissue-dependent three-dimensional patterns provide tissues with characteristic mechanical properties. Understanding the molecular composition of collagen fibrils and the interactions that define their specific tissue organization is, therefore, an essential aspect of molecular biomechanics.

Collagen fibrils are composed of subsets of fibrillar collagen types, and extensive data are available on the chemistry and molecular biology of the members of the fibrillar collagen family of proteins (types I, II, III, V, and XI) (Jacenko et al., 1991). Little is known, however, about the interactions that lead to establishment and maintenance of tissue-specific fibril patterns. It is likely that interactions involving fibril surfaces are important. Therefore, proteins that bind to fibrillar collagens are likely to play a role in organizing tissue-specific fibril scaffolds. Of particular interest are members of a recently discovered class of extracellular matrix molecules, named FACIT (fibril-associated collagens with interrupted triple-helices) molecules for their association with collagen fibrils and the presence of collagen-like, triple-helical domains in their structure (Jacenko et al., 1991). In fact, the multi-domain structure of FACIT molecules suggests that they represent macromolecular bridges between collagen fibrils and other matrix components and/or cells.

Several members of the FACIT family have been identified. These include types IX (van der Rest et al., 1985), XII (Gordon et al., 1989), XIV (Castagnola et al., 1992), XVI (Pan et al., 1992), and Y (Yoshioka et al., 1992) collagen. A characteristic identifying feature is the presence of a highly conserved triple-helical domain (called the COL1 domain) of about 100 amino acid residues (per chain) at the carboxyl end of the FACIT polypeptides (Shaw and Olsen, 1991). Their amino-terminal regions, however, are quite different. Type IX collagen, the first discovered member of the family, is found in cartilage and in some non-cartilage tissues such as vitreous humor. Type IX molecules contain α1(IX), α2(IX), and α3(IX) chains that form a long and a short triple-helical arm connected by a hinge region (Jacenko et al., 1991, see references within). The long arm is associated with the surface of type II-containing fibrils, while the short arm projects into the perifibrillar matrix (Vaughan et al., 1988). In cartilage, the α1(IX) chain encodes a globular amino-terminal domain, called NC4, located at the tip of the short arm. In the embryonic chick cornea and the vitreous, the NC4 is missing and replaced by a short, alternative amino acid sequence, because of the utilization of an alternative transcription start site in the α1(IX) collagen gene (Svoboda et al., 1988; Nishimura et al., 1989). It has been suggested, therefore, that the synthesis of two different forms of type IX represents a mechanism for altering the surface and interaction properties of type II-containing fibrils in a tissue-specific manner (Nishimura et al., 1989).

To examine this hypothesis we (Nakata et al., 1993) have analyzed the consequences of introducing a trans-dominant negative mutation in α1(IX) into transgenic mice. To generate the mutation, an α1(IX) cDNA, with a central in-frame deletion, was put under the control of the rat α1(II) collagen promoter/enhancer, previously demonstrated to confer tissue-specific expression in transgenic mice. The construct was injected into fertilized eggs, and two lines of transgenic mice were generated. In one founder, m8-1, about five copies of the transgene (per diploid genome) were integrated at a single site as a head-to-tail with head-to-head concatemer. The second founder, m22-1, contained about three copies as a head-to-tail concatemer at a different integration site. Western blots of cartilage extracts showed expression of the shortened α1(IX) chain; the ratio between this transgene product and the full-length endogenous α1(IX) chain was about 1:1 in homozygous offspring of the two lines. Mice homozygous for the transgene in both lines showed a mild proportionate dwarfism; heterozygous offspring showed no major skeletal defect.

Histological analysis of articular cartilage of knee-joints showed, however, early osteoarthritic changes in all offspring. At 4-6 months of age, there was erosion of the articular surface and loss of Safranin-0 staining. These changes progressed in both femoral and tibial condyles, leading to fissure formation and substantial loss of articular cartilage. Evaluation using a modified Mankin histological grading system, showed that offspring of both transgenic lines developed severe osteoarthritis with age while normal control mice did not.

Based on these findings we conclude that type IX collagen, while not essential for the developmental assembly of hyaline cartilage is important for the ability of articular cartilage to withstand mechanical stress. The early loss of proteoglycans (as judged by loss of Safranin 0-staining) in the cartilage of mutant mice is consistent with previous suggestions that type IX collagen molecules may represent bridges between collagen fibrils and proteoglycans (Shaw and Olsen, 1991).

The cartilage abnormalities in the transgenic mice that carry a truncated α1(IX) collagen gene are limited to articular cartilage; growth plate cartilage appears normal. This suggests that the role of type IX collagen in the growth plate is relatively unimportant. This is perhaps not very surprising since it has been demonstrated that expression of type IX collagen is significantly reduced in the growth plate as the chondrocytes undergo hypertrophy. The major collagen produced by hypertrophic chondrocytes in areas of endochondral ossification (EO) is a unique homotrimeric short-chain collagen, type X. The onset of type X collagen production in tissues undergoing EO coincides with the onset of hypertrophy; as hypertrophy progresses, type X collagen synthesis increases until the protein represents $\approx$45% of total collagen produced by hypertrophic cells (Reginato et al., 1986). The tissue-specific expression of type X collagen, in addition to certain biochemical properties of the protein, suggest an important role for type X during EO (Jacenko et al., 1991, see references within).

To investigate the function and regulation of type X collagen during EO, we generated transgenic mice with a dominant negative phenotype for type X. For this purpose, four different chicken type X DNA constructs encoding a truncated type X collagen subunit were made. The design of the constructs was as follows: the chicken type X collagen gene consists of three exons, encoding a 59 kd polypeptide, α1(X), composed of two globular domains which flank a central, triple-helical domain (LuValle et al., 1988). Assembly of three α1(X) polypeptide subunits into trimeric molecules presumably occurs through an interaction of the carboxyl-terminal domains, followed by folding of the triple-helix from the carboxyl towards the amino end (Brass et al., 1992). Our constructs contained portions of the 5' regulatory region of the chicken α1(X) gene (see below), and a cDNA with either a 21 or a 293 codon deletion in the triple-helical coding domain of type X collagen. Given the high degree of homology between the chicken and mouse α1(X) carboxyl domains (Apte et al., 1992), we expected the truncated chicken polypeptides to compete with the endogenous mouse α1(X) chains for assembly, and thus prevent proper triple-helical folding due to the deletion within the triple-helix. This would in turn cause a depletion of the endogenous, functional type X collagen. The 5' regulatory region of the chicken α1(X) gene was shown by transient transfection experiments to contain promoter and silencer elements that are active in both chicken and mammalian cells (LuValle et al., 1993). To confirm this *in vivo*, either 4700 bp or 1600 bp of this region were included in the constructs.

Microinjection of the above four constructs into pronuclei of fertilized mouse eggs yielded twenty-one founder mice, which where identified by genomic Southern blotting. Subsequent breeding established nineteen independent transgenic lines, each with different insertion sites of the transgene, as determined from Southern blots. Expression of the transgene was confirmed by Northern blot analysis, and by localization of the transgene protein to hypertrophic cartilage in newborn, genotypically positive mice. To date, thirteen of these lines, representing all four constructs, yielded mice with a similar phenotype.

Briefly, all transgenic mice were indistinguishable from their littermates at birth by visual inspection. However, about day 16/17 after birth, approximately 20% of the genotypically positive pups developed progressive hunching of the back, gradual hind limb paresis, wasting, and respiratory problems, and died by day 20-21. X-ray analysis, as well as whole mount Alizarin red staining of these mice, revealed pronounced lordosis of the cervical vertebral column and thoracolumbar kyphosis. Histology revealed that all tissues undergoing EO were affected similarly in genotypically positive mice, although the severity of the phenotype varied among the different transgenic lines. Analysis of the metaphyseal region of long bones and vertebrae showed alterations in three tissue zones: the growth plate, the zone of calcified trabeculae, and the bone marrow (the latter only in the 20% of the transgenics that developed the hunch-back phenotype).

Firstly, a compression of the growth plates was observed in the region of chondrocyte hypertrophy. The resting and proliferating chondrocytes often appeared unaffected, while hypertrophic chondrocytes were flattened as if the matrix around them had collapsed, and had pycnotic nuclei. The extent of hypertrophy was also not as extensive as that seen in genotypically negative mice. Secondly, the number and size of newly formed bony trabeculae, composed of calcified hypertrophic cartilage cores with newly deposited bone on the surface, was significantly reduced. It is noteworthy though that despite this reduction, all the trabeculae present (including periosteal bone), showed no abnormalities in the degree of mineralization, as seen by Alizarin red S staining. The third alteration, seen only in the mice that developed a hunch-back phenotype, involved the bone marrow. A predominance of mature erythrocytes, with a striking reduction of leukocytes was observed in the marrow. Analysis of the lymphatic organs of these mutant mice revealed a dramatically reduced thymus with a paucity of lymphocytes in the cortex region. The spleen was also reduced, and occasionally discolored.

The fact that the above phenotype was observed in 13 different transgenic lines, each with an independent integration site, suggests that the observed phenotype cannot be caused by disruption of an endogenous gene through transgene insertion. Furthermore, expression of the transgene in hypertrophic cartilage suggests that the chicken $\alpha 1(X)$ promoter and 5' regulatory elements are functional in mice, and that the transgene product colocalizes with areas showing the histological defect. It is therefore likely that the observed phenotype of spondylometaphyseal dysplasia in the transgenic mice is a direct result of the disruption of mouse type X collagen function.

The compression of growth plates in the zone of hypertrophic cartilage where type X is synthesized suggests that one function of type X collagen may be to provide a transient and additional structural support in a matrix undergoing degradation of the type II-type IX fibrillar scaffold and removal of proteoglycans. Type X may fulfill such a structural role by forming a hexagonal lattice in the pericellular space of hypertrophic chondrocytes seen *in vitro* (Kwan et al., 1991), similar to that described for type VIII collagen, the structural homologue of type X, in the Descemet's membrane (Sawada et al., 1989).

In summary, the observed defects in mice with a dominant negative phenotype for type X collagen indicate that type X collagen is required for normal skeletal morphogenesis, and also implicate COL10A1 as a candidate gene for certain human spondylometaphyseal dysplasias.

ACKNOWLEDGEMENTS

The research reviewed here was supported in part by grants AR36819 and AR36820 from the National Institutes of Health. Expert secretarial assistance was provided by Ms. M. Jakoulov.

REFERENCES

Apte SS, Seldin MF, Hayashi M and Olsen BR (1992): Cloning of the human and mouse type X collagen genes and mapping of the mouse type X collagen gene to chromosome 10. *Eur J Biochem* 206: 217-224.

Brass A, Kadler K, Thomas JT, Grant ME and Boot-Handford RP (1992): The fibrillar collagens, collagen VIII, collagen X, and the C1q complement proteins share a similar domain in their C-terminal non-collagenous regions. *FEBS Lett* 303: 126-128.

Castagnola P, Tavella S, Gerecke DR, Dublet B, Gordon MK, Seyer J, Cancedda R, van der Rest M and Olsen BR (1992): Tissue-specific expression of type XIV collagen -a member of the FACIT class of collagens. *Eur J Cell Biol* 59: 340-347.

Gordon MK, Gerecke DR, Dublet B, van der Rest M and Olsen BR (1989): Type XII collagen: A large multidomain molecule with partial homology to type IX collagen. *J Biol Chem* 264: 19772-19778.

Jacenko O, Olsen BR and LuValle P (1991): Organization and regulation of collagen genes. *Crit Rev Euk Gene Exp* 1: 327-353.

Kwan APL, Cummings CE, Chapman JA and Grant ME (1991): Macromolecular organization of chicken type X collagen *in vitro*. *J Cell Biol* 114: 597-604.

LuValle P, Ninomiya Y, Rosenblum ND and Olsen BR (1988): The type X collagen gene: Intron sequences split the 5' untranslated region and separate the coding regions for the non-collagenous amino-terminal and triple-helical domains. *J Biol Chem* 1988; 263:18378-18385.

LuValle P, Jacenko O, Iwamoto M, Pacifici M and Olsen BR (1993): From cartilage to bone -the role of collagenous proteins. In: *Molecular Basis of Morphogenesis*, 51st Annual Symposium of the Society for Developmental Biology, Bernfield M, ed. New York: John Wiley & Sons, Wiley-Liss Division, in press.

Nakata K, Ono K, Miyazaki J-I, Olsen BR, Muragaki Y, Adachi E, Yamamura K-I and Kimura T (1993): Osteoarthritis associated with mild chondrodysplasia in transgenic mice expressing α1(IX) collagen chains with a central deletion. *Proc Natl Acad Sci USA*, in press.

Nishimura I, Muragaki Y and Olsen BR (1989): Tissue-specific forms of type IX collagen-proteoglycan arise from the use of two widely separated promoters. *J Biol Chem* 264: 20033-20041.

Pan T-C, Zhang R-Z, Mattei M-G, Timpl R and Chu M-L (1992): Cloning and chromosomal location of human α1(XVI) collagen. *Proc Natl Acad Sci USA* 89: 6565-6569.

Reginato A, Lash J and Jimenez S (1986): Biosynthetic expression of type X collagen in embryonic chick sternum cartilage during development. *J Biol Chem* 261: 2897-2903.

Sawada H, Konomi H and Hirosawa K (1990): Characterization of the collagen in the hexagonal lattice of Descemet's membrane: its relation to type VIII collagen. *J Cell Biol* 110: 219-227.

Svoboda KK, Nishimura I, Sugrue SP, Ninomiya Y and Olsen BR (1988): Embryonic chicken cornea and cartilage synthesize type IX collagen molecules with different amino-terminal domains. *Proc Natl Acad Sci USA* 85: 7496-7500.

van der Rest M, Mayne R, Ninomiya Y, Seidah NG, Chretien M and Olsen BR (1985): The structure of type IX collagen. *J Biol Chem* 260: 220-225.

Vaughan L, Mendler M, Huber S, Bruckner P, Winterhalter KH, Irwin MH and Mayne R (1988): D-periodic distribution of collagen type IX along cartilage fibers. *J Cell Biol* 106: 991-997.

Yoshioka H, Zhang H, Ramirez F, Mattei M-G, Moradi-Ameli M, van der Rest M and Gordon MK (1992): Synteny between the loci for a novel FACIT-like collagen locus (D6S228E) and α1(IX) collagen (COL9A1) on 6q12-q14 in humans. *Genomics* 13: 884-886.

ELASTIC FIBER ORGANIZATION

Elaine C. Davis
Department of Cell Biology and Physiology
Washington University School of Medicine
St. Louis, MO

Robert P. Mecham
Departments of Cell Biology and Medicine
Washington University Medical Center
St. Louis, MO

INTRODUCTION

Elastic fibers are components of virtually all mammalian connective tissues. Elastin, the major protein component of elastic fibers, has unique elastomeric properties which provide the reversible deformability that is critical to arterial vessels, lungs and skin. Without such a property, the structural integrity and function of these tissues would be greatly impaired.

The distribution and organization of elastic fibers in various tissues and organs can be demonstrated at the light microscope level by characteristic staining reactions which date back to the turn of the century. In ligamentum nuchae, interwoven rope-like fibers are observed which branch and rejoin in a three-dimensional network. Elastic fibers of the lung appear similarly arranged, however, without the longitudinal orientation of ligament fibers. In the aorta, elastin is in the form of concentric sheets or lamellae. A combination of these two forms is seen in the skin, flattened bands of elastin in the dermis and fine filamentous networks in the papillary layer. A third form of elastin is apparent in elastic cartilage where large anastomosing fibers form a three-dimensional honeycomb pattern.

Electron microscopy reveals elastic fibers to consist of two morphologically distinct components (Fig.1): an amorphous core of insoluble elastin and a peripheral mantle of 10 - 12 nm microfibrils (Low, 1962; Greenlee et al., 1966; Ross and Bornstein, 1969). Studies on elastic fiber formation have shown that microfibrils appear first in the extracellular matrix prior to the deposition of elastin. Additional studies of elastic fiber formation, both in vivo and in vitro, corroborate these findings, in that, microfibrils are initially deposited in the extracellular matrix and subsequently become infiltrated with elastin deposits which ultimately merge to form mature elastic fibers (Greenlee et al., 1966; Ross, 1971)

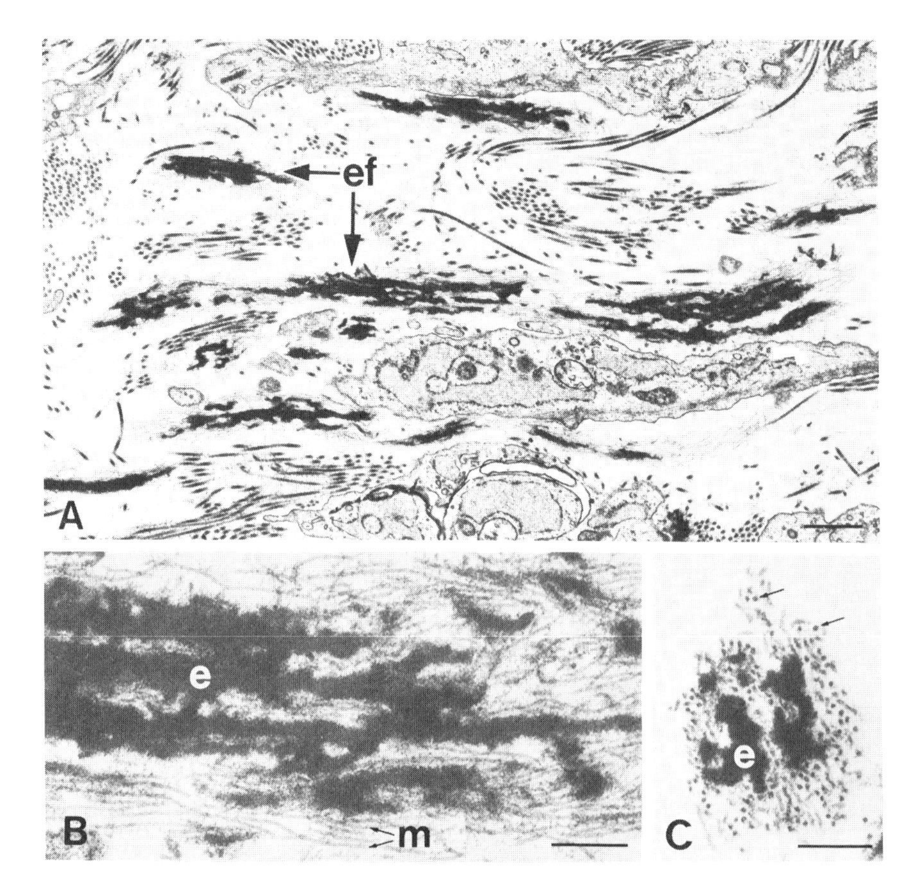

FIGURE 1. Electron micrograph of developing elastic fibers (ef) in bovine ligamentum nuchae (**A**). Longitudinal section (**B**) of an elastic fiber showing the 10 nm microfibrils (m) and elastin (e). In cross-section (**C**), the micrifibrils have a tubular appearance (arrows). Bars: (A) 1 µm; (B,C) 0.25 µm.

ELASTIC FIBER COMPOSITION

1. Elastin

Elastin is one of the most insoluble protein in the body and is extremely resilient to the standard denaturation and degradation techniques usually employed to isolate proteins. For this reason it has been one of the more difficult proteins to study. Its stability in harsh isolation procedures, however, has been utilized as a basis for successful methods of elastin purification.

More than 99% of the total amount of elastin present in normal tissue is in a mature form; a highly cross-linked insoluble network (Rucker and Tinker, 1977). This mature insoluble elastin is extremely hydrophobic due to a high content of nonpolar amino acids such as glycine, alanine, valine and proline. Valine and proline account for approximately 14% and 12% of the amino acids, respectively. One-third of the residues are glycine and almost one-quarter are alanine (data in Rucker and Tinker, 1977). The polar amino acids aspartate, glutamate, lysine and arginine comprise only 5% of the residues of the insoluble elastic fiber. In contrast, the soluble elastin precursor contains over 40 lysine residues per thousand total residues. The difference in lysine content is due to the conversion of lysine residues into the cross-links desmosine and isodesmosine; a process mediated by the copper-dependent enzyme, lysyl oxidase (Kagan and Trackman, 1991).

2. *Microfibrils*

The composition of microfibrils has proven difficult to determine due to their insolubility and the extent to which they associate with elastin and other extracellular matrix components. Increasing evidence suggests that microfibrils consist of a number a different protein assembled into a complex structure that is critical for elastic fiber assembly. Two of the best characterized glycoproteins that appear to be constituents of microfibrils are fibrillin (Sakai et al., 1986) and microfibril-associated glycoprotein (MAGP) (Gibson et al., 1986).

a) Fibrillin: Fibrillin is a large (350 kDa) glycoprotein that is a component of all microfibrils. The fibrillin molecule is made up of several cysteine-rich sequences that have homology to the precursor of epidermal growth factor (EGF) and TGF-ß1 binding protein (Maslen et al., 1991). Recent evidence suggests that the fibrillins are actually a gene family, with to date, three and possibly more members. Interestingly, mutations in the fibrillin gene on human chromosome 15 (Fib-15) have been linked to Marfan syndrome (Dietz et al., 1991; McKusick, 1991) and a Marfan-like syndrome, congenital contractural arachnodactyly, has been linked to mutations in a second fibrillin gene on chromosome 5 (Fib-5) (Lee et al., 1991). Fib-5 shares a high degree of homology to Fib-15. Both contain EGF-like repeats with the consensus sequence for hydroxylation of asparagine or aspartic acid and one RGD sequence. Recently, a third distinct fibrillin has been identified, called fibrillin-like protein (FLP). On comparison with the other fibrillin proteins, FLP demonstrates the same motif structure but differs in sequence within the individual motifs (manuscript in preparation). The genomic heterogeneity of the fibrillin proteins thus suggests that microfibrils may consist of a family of morphologically and immunologically similar filaments rather than a single entity (Fig. 2).

b) MAGP: MAGP is a 32 kDa protein that contains two structurally distinct regions: an amino-terminal domain, rich in glutamine, proline and acidic amino acids; and a carboxyl-terminal domain, containing all 13 of the cysteine residues and most of the basic amino acids (Gibson et al., 1991). In contrast to Fib-5 and Fib-15, MAGP does not contain an RGD sequence. Immunofluorescence localization of MAGP with affinity purified antibodies, showed the distribution of the glycoprotein to correspond to that of elastin-associated microfibrils and microfibril bundles free of elastin in skin, periodontal ligament and ciliary zonules (Gibson and Cleary, 1987). The specificity of the anti-MAGP antibody for microfibrils in a wide range of tissues was further demonstrated by immunogold labeling at the electron microscope level (Kumaratilake et al., 1989). Using cDNA cloning, MAGP was determined to be a distinct glycoprotein and not derived from fibrillin (Gibson et al., 1991).

c) Other proteins: Gibson and colleagues (1989) obtained 78, 70 and 25 kDa proteins from the ligamentum nuchae extract. The 78 kDa protein was demonstrated by immunogold electron microscopy with an affinity purified

Domain Alignment of Three Fibrillin Proteins
(Fib-15 and Fib-5 based on published sequences)

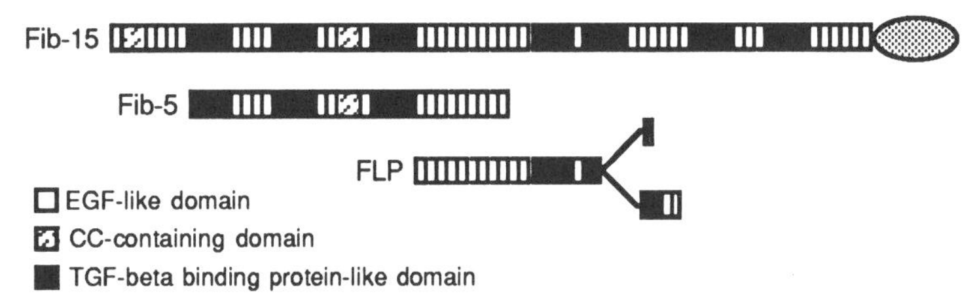

FIGURE 2. Structural domains of the human fibrillins, Fib-15 and Fib-5, and the bovine fibrillin, fibrillin-like protein (FLP).

antibody to bind specifically to elastin-associated microfibrils. Similarly, the 25 kDa protein also appears to be a microfibrillar component, while the origin of the 70 kDa protein has yet to be determined (Gibson et al., 1991). In addition to these proteins, several other components of reductive guanidine-hydrochloride extracts have also been isolated from elastin-rich tissues and used to produce antisera that label elastin-associated microfibrils. These components include 32 kDa and 250 kDa proteins isolated from bovine ciliary zonular fibrils (Streeten and Gibson, 1988), a 70 kDa protein also isolated from bovine zonular fibrils (Mecham et al., 1988) and a 35 kDa protein isolated from bovine ligamentum nuchae that shows amine oxidase activity and aggregates to form 11 nm fibrils (Serafini-Fracassini et al., 1981; Jaques and Serafini-Fracassini, 1986). Recently, a new protein, termed associated microfibril protein (AMP), has been identified that appears to be a 58 kDa protein that is post-translationally modified into a 32 kDa fragment which localizes with microfibrils (Horrigan et al., 1992). The relationship of these proteins to fibrillin and MAGP, and their contribution to the fundamental structure of microfibrils, remains to be elucidated.

Microfibrils also appear to be associated with a number of different proteins including fibronectin, amyloid P component, vitronectin and lysyl oxidase. The microfibril-associated proteins often appear to be selectively distributed on microfibrils dependent upon the tissue location. Although this suggests that some constituent(s) of microfibrils may differ in different regions, the widespread distribution of fibrillin (Sakai et al., 1986) and MAGP (Gibson et al., 1986) suggests that at least some constituents may be ubiquitous.

ASSEMBLY OF INSOLUBLE ELASTIN

Critical to the elastomeric properties of elastin is the crosslinking of soluble elastin precursor molecules to each other. The initial step in this process is the deamination of the ε-amino group of lysine side chains in tropoelastin by the copper-requiring enzyme lysyl oxidase. The reactive aldehyde that is formed (α-amino adipic δ-semi-aldehyde, or allysine) can then condense with a second aldehyde residue to form allysine aldol or with an ε-amino group on lysine to form dehydrolysinonorleucine. Allysine aldol and dehydrolysinonorleucine can then condense to form the pyridinium cross-links called desmosine and isodesmosine, both of which are unique to vertebrate elastin and can be used as distinctive markers for the protein. It is estimated that all but 5 of the 38 lysine residues of bovine tropoelastin are modified during elastin maturation. In bovine ligamentum nuchae elastin, for example, there is estimated to be 2 desmosines, 1 isodesmosine, 2 lysinonorleucines and 4 aldol condensation products.

The precise physico-chemical properties that account for the rubber-like characteristics of elastin have not been fully characterized. Biophysical studies of purified elastic fibers have suggested that elastin behaves as a classical "rubber" modeled after a cross-linked network of randomly oriented chains that lack both short- and long-range order (Hoeve, 1974). According to the rubber theory, an unstretched rubber network is in a state of maximum disorder and, hence, maximum entropy. Stretching the rubber induces order in the direction of extension, thereby decreasing entropy (crosslinking is important for transferring the stress throughout the polymer). When the external force is removed, elastic recoil will occur because the partially ordered chains return spontaneously to their initial random state in order to return to a level of maximum entropy. Elastin exhibits the physical features characteristic of a polymeric rubber, including interchain cross-linkages and chains that behave as a kinetically free, random chain network.

While the dominant pattern of elastin is that of a random structure, the elastic fiber is more complicated than a simple-three dimensional array of random chains. Several physico-chemical studies suggest that there are regions of the molecule that exhibit areas of local order. For example, the alanine-rich sequences that define cross-linking domains have been shown to be in an extended helix conformation and Urry and coworkers (1983)have shown extensive secondary structure (e.g. beta-turns giving rise to beta-spirals) associated with several repeating hydrophobic sequences. These microdomains of secondary order could serve to restrict movement around chemical bonds, and thus the rotational freedom assumed for an ideal rubber may not always hold for all areas of elastin.

FORMATION OF ELASTIC FIBERS

Although the appearance of elastic fibers can be readily observed during the development of elastogenic tissues, remarkably little is known concerning the events involved in their formation. Both the intracellular trafficking of microfibril proteins and tropoelastin, and their functional association in the extracellular matrix remain unclear.

The presence of microfibrils precedes that of elastin in the extracellular space during early embryonic development of elastic fibers (Fahrenbach et al., 1966; Greenlee et al., 1966). These microfibrils are organized into small bundles that frequently form a close association with the cell surface. The cell surface in this region often shows the presence of a membrane-associated dense plaque with an abundance of microfilaments in the adjacent cytoplasm (Fig. 3). Interestingly, actin was copurified with several components of an elastin receptor complex

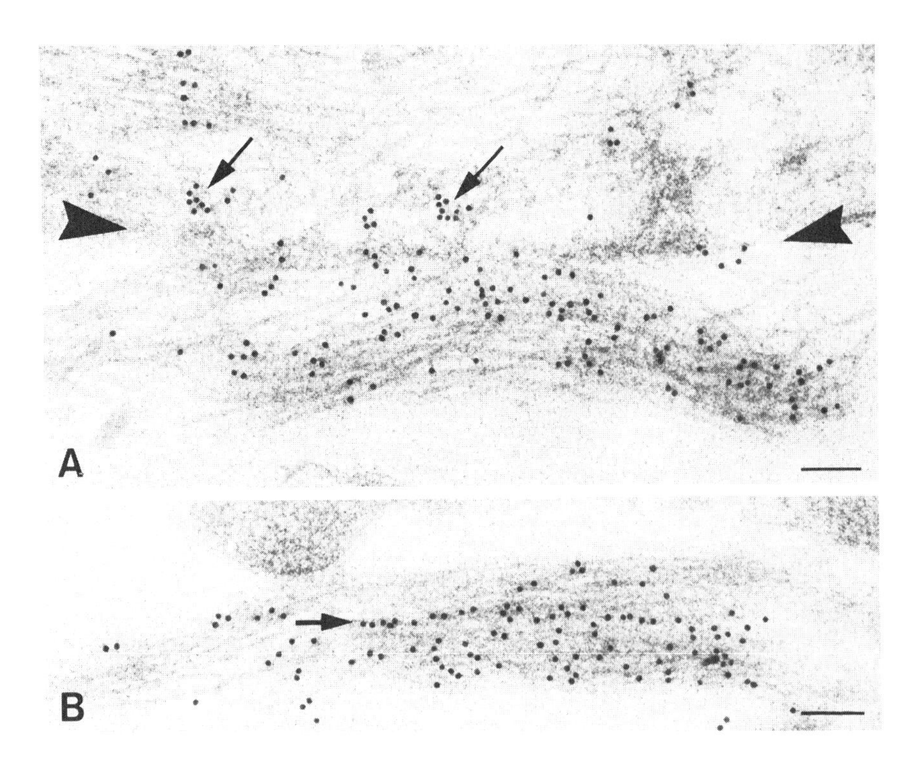

FIGURE 3. Immunolocalization of tropoelastin on developing elastic fibers in bovine ligamentum nuchae. Intracellular labeling for tropoelastin (arrows) is seen in a region of the cell where an elastic fiber forms a close association with the fibroblast cell membrane (arrowheads) (**A**). A developing elastic fiber shows a periodic localization of tropoelastin molecules along the microfibrils (arrow) (**B**). Gold particles are 10 nm in diameter. Bars: (A,B) 100 nm.

found on the membrane of elastin-producing cells (Mecham et al., 1989). The affinity of the receptor for elastin and the distribution of the receptor on the cell surface appears to be influenced by the underlying cytoskeleton (Mecham et al., 1991a). In addition to the elastin receptor, a filamentous distribution of intracellular lysyl oxidase suggests that this enzyme is also associated with cytoskeletal proteins (Waksaki and Ooshima, 1990). Thus, several lines of evidence now suggest that the cytoskeleton may play a critical role in elastic fiber assembly.

Although a functional relationship between the elastin producing cell and developing elastic fiber remains to be established, the present observations and early electron micrographs showing elastic fibers forming within folds or crevices of the plasma membrane suggest that elastogenesis requires intimate contact between the plasma membrane of the elastogenic cell and the surface of the developing elastic fiber. It is possible that the tropoelastin monomers must be presented on the cell surface in such an orientation to allow for an alignment on the microfibril that is conducive to crosslink formation. The synthesis of microfibril proteins decreases during development prior to that of elastin and the mature elastic fiber has only a sparse mantle of microfibrils. Thus, following initial deposition of correctly aligned tropoelastin molecules, subsequent monomers may be able to assemble on the fiber in a more independent manner.

DEGRADATION AND TURNOVER OF ELASTIN

Physiological turnover of insoluble elastin appears to be very slow, with a half-life approaching that of an animals' life-time (Lefevre and Rucker, 1980; Shapiro et al., 1991). Thus, under normal conditions, very little remodeling of elastic fibers occurs in adult life. While tropoelastin has been shown to be sensitive to proteolytic degradation before cross-linking into insoluble elastin, there is no evidence that significant intracellular degradation occurs prior to secretion. In contrast to collagen synthesis, where a large percentage of newly synthesized interstitial collagen undergoes intracellular catabolism, studies in cultured cells suggest that only about 1% of newly synthesized elastin is degraded per hour.

Increased elastin destruction does occur, however, in certain pathological conditions. This is usually a result of either the release of powerful elastinolytic serine and metallo-proteases from inflammatory cells of bacteria, or because of a genetic deficiency in the naturally occurring elastase inhibitor α1–antitrypsin. Although elastin synthesis in the adult is negligible, some pathological conditions result in reactivation of elastin synthesis. While this suggests the potential for elastic fiber repair following injury, the cells that produce the elastin are often phenotypically altered cells that produce different connective tissue components than do normal cells (Mecham et al., 1991b). Furthermore, the elastin that is produced in response to injury is often morphologically disorganized and may be inappropriate for, or even impair, normal function (Kuhn et al., 1976; Rucker and Dubick, 1984).

SUMMARY

The organization of elastic fibers involves a complex series of events including the intracellular synthesis and secretion of tropoelastin, microfibril proteins and lysyl oxidase and their coordinated assembly in the extracellular matrix. Although a great deal of questions remain to be answered, recent advancements in cell and molecular biology, as well as model systems to study elastogenesis, will soon provide insight into the mechanism of normal elastic fiber assembly and that which occurs under pathological conditions.

ACKNOWLEDGMENTS

The research from our laboratory discussed in this review was supported in part by grants to R.P. Mecham from the National Institutes of Health (HL26499, HL41926). E.C. Davis is a Heart and Stroke Foundation of Canada Research Fellow.

REFERENCES

Dietz HC, Cutting GR, Pyeritz RE, Maslen CL, Sakai LY, Corson GM, EG, Hamosh A, Nanthakumar EJ, Curristin SM, Stetten G, Meyers DA and Francomano CA (1991): Marfan syndrome caused by a recurrent de novo missense mutation in the fibrillin gene. *Nature (Lond)* 352: 337-339.

Fahrenbach WH, Sandberg LB and Cleary EG (1966): Ultrastructural studies on early elastogenesis. *Anat Rec* 155: 563-576.

Gibson MA and Cleary EG (1987): The immunohistochemical localization of microfibril-associated glycoprotein (MAGP) in elastic and non-elastic tissues. *Immunol Cell Biol* 65: 345-356.

Gibson MA, Hughes JL, Fanning JC and Cleary EG (1986): The major antigen of elastin-associated microfibrils is a 31 kDa glycoprotein. *J Biol Chem* 261: 11429-11436.

Gibson MA, Kumaratilake JS and Cleary EG (1989): The protein components of the 12-nanometer microfibrils of elastic and non-elastic tissues. *J Biol Chem* 264: 4590-4598.

Gibson MA, Sandberg LB, Grosso LE and Cleary EG (1991): Complementary DNA cloning establishes microfibril-associated glycoprotein (MAGP) to be a discrete component of the elastin-associated microfibrils. *J Biol Chem* 266: 7596-7601.

Greenlee TKJ, Ross R and Hartman JL (1966): The fine structure of elastic fibers. *J Cell Biol* 30: 59-71.

Hoeve CAJ (1974): The elastic properties of elastin. *Biopolymers* 13: 677-686.

Horrigan SK, Rich CB, Streeten BW, Li ZY and Foster JA (1992): Characterization of an associated microfibril protein through recombinant DNA techniques. *J Biol Chem* 267: 10087-10095.

Jaques A and Serafini-Fracassini A (1986): Immunolocalisation of a 35K structural glycoprotein to elastin-associated microfibrils. *J Ultrastruct Mol Struct Res* 95: 218-227.

Kagan HM and Trackman PC (1991): Properties and function of lysyl oxidase. *Am J Respir Cell Molec Biol* 5: 206-210.

Kuhn CI, Yu SY, Chraplyvy M, Linder HE and Senior RM (1976): The induction of emphysema with elastase. II. Changes in connective tissue. *Lab Invest* 34: 372-380.

Kumaratilake JS, Gibson MA, Fanning JC and Cleary EG (1989): The tissue distribution of microfibrils reacting with a monospecific antibody to MAGP, the major glycoprotein antigen of elastin-associated microfibrils. *Eur J Cell Biol* 50: 117-127.

Lee B, Goodfrey M, Vitale E, Hori H, Mattei M-G, Sarfarazi M, Tsipouras P, Ramirez F and Hollister DW (1991): Linkage of Marfan syndrome and a phenotypically related disorder to two different fibrillin genes. *Nature (Lond)* 352: 330-334.

Lefevre M and Rucker RB (1980): Aorta elastin turnover in normal and hypercholesterolemic Japanese quail. *Biochim Biophys Acta* 630: 519-529.

Low FN (1962): Microfibrils: Fine filamentous components of the tissue space. *Anat Rec* 142: 131-137.

Maslen CL, Corson GM, Maddox BK, Glanville RW and Sakai LY (1991): Partial sequence of a candidate gene for the Marfan syndrome. *Nature (Lond)* 352: 334-337.
McKusick VA (1991): The defect in Marfan syndrome. *Nature (Lond)* 352: 279-281.
Mecham RP, Hinek A, Cleary EG, Kucich U, Lee SJ and Rosenbloom J (1988): Development of immunoreagents to ciliary zonules that react with protein components of elastic fiber microfibrils and with elastin-producing cells. *Biochem Biophys Res Commun* 151: 822-826.
Mecham RP, Hinek A, Entwistle R, Wrenn DS, Griffin GL and Senior RM (1989): Elastin binds to a multifunctional 67 kD peripheral membrane protein. *Biochemistry* 28: 3716-3722.
Mecham RP, Whitehouse L, Hay M, Hinek A and Sheetz M (1991a): Ligand affinity of the 67 kDa elastin/laminin binding protein is modulated by the protein's lectin domain: Visualization of elastin/laminin receptor complexes with gold-tagged ligands. *J Cell Biol* 113: 187-194.
Mecham RP, Stenmark KR and Parks WC (1991b): Connective tissue production by vascular smooth muscle in development and disease. *Chest* 99: 43S-47S.
Ross R (1971): The smooth muscle cell. II. Growth of smooth muscle in culture and formation of elastic fibers. *J Cell Biol* 50: 172-186.
Ross R and Bornstein P (1969): The elastic fiber. I. The separation and partial characterization of its macromolecular components. *J Cell Biol* 40: 366-381.
Rucker RB and Dubick MA (1984): Elastin metabolism and chemistry: potential roles in lung development and structure. *Environ Health Perspect* 53: 179-191.
Rucker RB and Tinker D (1977): Structure and metabolism of arterial elastin. *Int Rev Exp Pathol* 17: 1-47.
Sakai LY, Keene DR and Engvall E (1986): Fibrillin, a new 350-kD glycoprotein, is a component of extracellular microfibrils. *J Cell Biol* 103: 2499-2509.
Serafini-Fracassini A, Ventrella G, Field MJ, Hinnie J, Onyezili NI and Griffiths R (1981): Characterization of a structural glycoprotein from bovine ligamentum nuchae exhibiting dual amine oxidase activity. *Biochemistry* 20: 5424-5429.
Shapiro SD, Endicott SK, Province MA, Pierce JA and Campbell EJ (1991): Marked longevity of human lung parenchymal elastic fibers deduced from prevalence of D-aspartate and nuclear weapons-related radiocarbon. *J Clin Invest* 87: 1828-1834.
Streeten BW and Gibson SA (1988): Identification of extractable proteins from the bovine ocular zonule: Major zonular antigens of 32 kD and 250 kD. *Current Eye Res* 7: 139-146.
Urry DW (1983): What is elastin; what is not. *Ultrastruct Pathol* 4: 227-251.
Waksaki H and Ooshima A (1990): Immunohistochemical localization of lysyl oxidase with monoclonal antibodies. *Lab Invest* 63: 377-384.

Section III

Use of Cultured Cells and the Potential of Stem Cells for Tissue Restoration

MYOBLAST MEDIATED GENE THERAPY

Helen M. Blau, Grace K. Pavlath and Jyotsna Dhawan
Department of Pharmacology
Stanford University Medical School
Stanford, CA 94305

ABSTRACT

A novel approach to drug delivery in the treatment of disease involves using cells to introduce genes into the body that express therapeutic proteins continuously (Friedman, 1989; Miller, 1990; Anderson, 1992; Miller, 1992). Myoblasts appear to be well suited for this purpose due to their unique biological properties. By contrast with other cell types, myoblasts become an integral part of the tissues into which they are injected. As a result, myoblasts are currently primary candidates as cellular vehicles for gene delivery in the treatment of both muscle and nonmuscle disorders. Transplanted myoblasts may be used to correct defects in well characterized inherited myopathies such as Duchenne muscular dystrophy (DMD) (Gussoni et al., 1992). In addition, myoblasts may provide genes to correct myopathies that are not understood at the molecular level. Finally, and perhaps most exciting, is the finding that genetically engineered myoblasts can be used to deliver nonmuscle recombinant proteins to the circulation (Dhawan et al., 1991; Barr and Leiden, 1991). Candidates for delivery include hormones, coagulation factors, and antitumor agents that could have broad applications ranging from the treatment of inherited hormone deficiencies to symptoms of aging, hemophilia, and cancer.

BACKGROUND

Cell-mediated gene therapy for the delivery of recombinant proteins to the circulation involves genetically engineering cells in tissue culture to produce a therapeutic protein product. The cells are extensively characterized in vitro and in animal models to ensure that the foreign genes are expressed, and that the cells are non-tumorigenic prior to injection into patients.

Gene therapy appears to have several advantages as a method of drug delivery. Problems of impurity of proteins that have caused AIDS in hemophiliacs, high costs of recombinant proteins which must be manufactured under quality control in bulk, and the cumbersome administration of these proteins by frequent injections could be circumvented by cell-mediated delivery. Of greatest significance, cell-mediated drug delivery could overcome problems associated with high dose therapy which is necessary when a protein with a short half life such as growth hormone (4 min in serum) or GM-CSF (35 min in serum) is administered by injection. If degradation rates are high, within a 24 hour period, serum levels of a therapeutic protein will range from exceedingly high to levels that are inadequate. As a result, early after injection there may be a risk of toxicity, later a risk of inefficacy. Thus, a major attraction of cell-mediated delivery is that proteins could be provided constitutively at relatively low, physiological levels.

A number of cell types have been tested as potential vehicles for systemic delivery of recombinant proteins, including keratinocytes and fibroblasts. Although longterm stable secretion of genetically engineered products was reported in some cases using established cell lines, this persistence of expression appeared to be due to tumor formation (Garver et al., 1987; Selden et al., 1987; Teumer et al., 1990; Scharfmann et al., 1991). In other cases, the gene product was not detected in serum either due to extinction of vector expression, antibody degradation of the protein, or lack of access to the circulation (Morgan et al., 1987; Selden et al., 1987; St. Louis and Verma, 1988; Teumer et al., 1990; Palmer et al., 1991; Scharfmann et al., 1991).

Studies of Muscle Patterning Reveal Unique Properties of Myoblasts

For the past decade we have studied the cell biology of muscle precursor cells and the patterning of muscle tissues in the course of normal and dystrophic development (Blau and Webster, 1981; Blau et al., 1983b; Webster et al., 1986; Webster et al., 1988b). These studies suggested that myoblasts might be particularly advantageous as vehicles for gene therapy. Myoblasts have the unusual capability of crossing basal lamina and disseminating within the tissue (Hughes and Blau, 1990). Although encased in a sheath of extracellular matrix in close apposition to a single fiber, progeny of endogenous "satellite" myoblasts (Mauro, 1961) fuse randomly with all muscle fibers in their vicinity (Hughes and Blau, 1992). In addition, myoblasts isolated from muscle tissue, grown in culture and injected back into muscle gain access to and become incorporated into mature muscle fibers of the host, both in normal (Hughes and Blau, 1992) and in dystrophic animals (Karpati et al., 1989; Partridge et al., 1989; Pavlath and Blau, unpublished). Thus, injected myoblasts contribute to a syncytium that is appropriately vascularized and innervated. By contrast, transplanted endothelial cells must be implanted on a matrix, fibroblasts require a source of FGF for sustenance, and primary keratinocytes lose access to the circulation. The longterm expression and secretion of the product of the engineered gene by injected myoblasts is likely to be a direct result of the unique biological features of the muscle cell.

Primary Human Myoblasts are Readily Isolated, Purified and Grown to Large Numbers

Another distinct advantage of muscle over other cell types is that human primary myoblasts can be readily isolated, purified, grown to large numbers and genetically engineered without losing their potential to differentiate. Primary human myoblasts can be obtained in abundance from biopsy or autopsy material (Blau and Webster 1981; Blau et al., 1982; Blau et al., 1983a,b; Costa et al., 1986; Kaplan and Blau, 1986; Shimizu et al., 1986; Webster et al., 1986; Gunning et al., 1987; Ham et al., 1988; Miller et al., 1988; Webster et al., 1988a,b; Kaplan et al., 1990; Webster and Blau, 1990; Mantegazza et al., 1991. Upon plating in culture, myoblasts exhibit an extensive proliferative capacity yielding between 10^{12} to 10^{18} cells per cell depending on the age of the donor (Webster and Blau, 1990). This represents a potential yield of kilograms of cells per 0.1 gm biopsy. These findings suggest that large numbers of myoblasts can easily be generated for use in myoblast-mediated gene therapy.

Human myoblasts are easily separated from other cell types present in dissociated muscle tissue with the fluorescence activated cell sorter using monoclonal antibodies specific to a muscle form of neural cell adhesion molecule (Figure 1; Webster et al., 1988a). Cells isolated from a post-natal biopsy that exhibited an initial composition of 79% myoblasts, were enriched to 99%, a purity that was maintained when the cells were re-analyzed three weeks later. Such primary human myoblasts

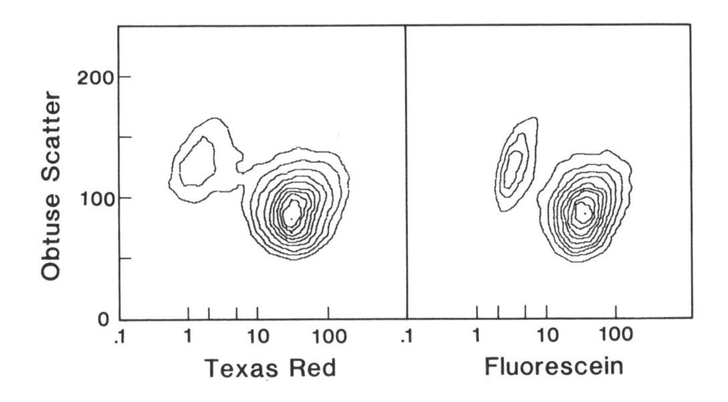

FIGURE 1. Efficient separation of myoblasts from fibroblasts can be achieved using an antibody against neural cell adhesion molecule (NCAM) detected by labeling with either Texas red or fluorescein. The same mixed-cell population is shown labeled with Texas red (A) or fluorescein (B) in two-dimensional contour plots of red or green immunofluorescence (abscissa) and intrinsic obtuse angle light scatter (ordinate), where 10% of the cells lie between each contour line. In this case, 87% of the cells were myoblasts (right-hand plots). Reprinted with permission (Webster et al., 1988a).

can be engineered in vitro to express foreign genes (Pavlath and Blau, unpublished observations).

Other advantages of myoblasts are that removal of muscle tissue from the donor and the return of engineered cells to the host are simple and relatively non-invasive manipulations. In addition, malignant transformation of human muscle cells in vitro has never been reported and the cells invariably senesce after a given number of doublings (Webster and Blau, 1990), suggesting that primary myogenic cells have low tumorigenicity. After fusion, implanted myoblast nuclei become post-mitotic, and in the mouse, have been shown to remain viable within the fibers for periods of at least 6 months (Hughes and Blau, unpublished observations). Thus, transplanted myoblasts persist and exhibit stable gene expression.

Phase I Clinical Trial of Myoblast Therapy in Duchenne Muscular Dystrophy Patients

The ultimate goal of cell therapy in DMD is to provide dystrophin, the protein missing in DMD that leads to progressive muscle degeneration and death around the second decade of life. Although it is unlikely that a cure for this particular disease will be achieved by myoblast transplantation, since skeletal, heart and diaphragm muscles are affected, an improvement in the quality of life may be possible. In a recent clinical trial, we transplanted normal myoblasts from a father or an unaffected sibling into the muscle of eight boys with DMD and assayed for the production of dystrophin (Gussoni et al., 1992). Three patients with deletions in the dystrophin gene expressed normal dystrophin transcripts in muscle biopsy specimens taken from the transplant site one month after myoblast injection and in one case, 6 months post transplant. No deleterious effects were observed in any of the patients. Using the polymerase chain reaction (PCR) we determined that the dystrophin in these biopsies derived from donor myoblast DNA. The use of PCR instead of immunofluorescence allowed a distinction between dystrophin expressed as a result of implantation of donor myoblasts or genetic reversion in host muscle cells leading to production of a truncated protein (Hoffman et al., 1990).

The expression of dystrophin in human muscle was surprisingly low by contrast with prior experience by us and by others with the mdx mouse model of DMD (Karpati et al., 1989; Partridge et al., 1989; Pavlath and Blau, unpublished observations). This observation shows that a direct extrapolation from animal to man is not always achieved. Indeed, there is increasing evidence that as in the case of DMD, animal models with the same genetic defect as their human counterparts do not fully recapitulate the human disease, limiting their utility as test systems for therapeutic strategies (Leiter et al., 1987). The basis for the low efficiency of myoblast transplantation in dystrophic human muscles is unknown. The problem could derive from the biology of the human disease or from technical aspects of delivery of cells into human muscles. To distinguish between these possibilities, a method is required for marking and tracking injected myogenic cells in humans. Furthermore, tests are needed in human muscle that does not suffer from infiltration by adipose and connective tissue cells.

Systemic Delivery of Recombinant Proteins by Injected Myoblasts

We and others (Barr and Leiden, 1991; Dhawan et al., 1991; Roman et al., 1992; Yao and Kurachi, 1992) tested whether genetically

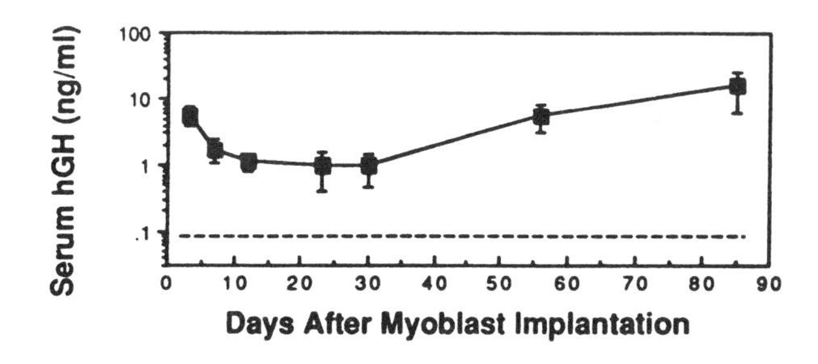

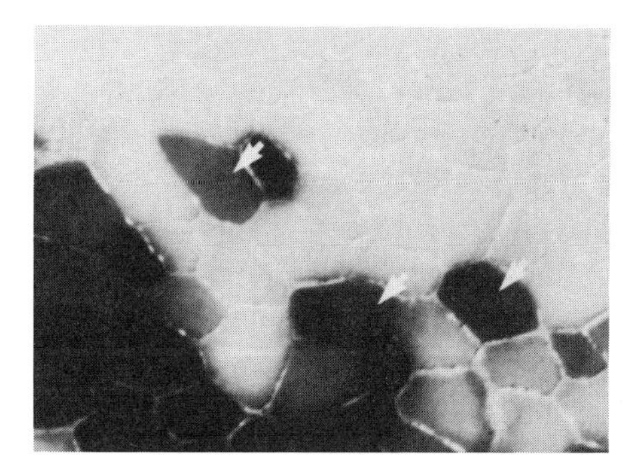

FIGURE 2 (Top) Persistent expression of hGH by virus-transduced myoblasts implanted in mouse muscle in vivo. A pool of transduced myoblasts was implanted into 24 C3H mice and serum hGH levels were monitored for 85 days by collection of tail blood for radioimmunoassay. Greater than 90% of the cells expressed and secreted hGH and 30% expressed β-gal, determined by clonal analysis in vitro. Each point represents the mean ± SD for 4 to 24 mice. The dashed line is the mean ± SD (0.08 ± 0.08 ng/ml) for serum samples from five uninjected control mice. (Bottom) Genetically engineered myoblasts can contribute to existing muscle fibers. Transverse sections of mouse legs 12 days after myoblast implantation show that β-galactosidase labelled myoblasts have fused into large diameter myofibers (arrows). Reprinted with permission (Dhawan et al., 1991).

engineered myoblasts could serve as vehicles for the systemic delivery of recombinant non-muscle proteins. A recombinant gene encoding human growth hormone (hGH) was stably introduced into cultured myoblasts with a retroviral vector. hGH was chosen for study because it has a very short half life in mouse serum (4 min) providing a stringent test for continuous production and access to the circulation over time. In addition, sensitive assays distinguish the mouse and human hGH proteins.

After injection of genetically engineered myoblasts into mouse muscle, hGH was detected in serum for 3 months (Figure 2, top). The fate of the injected myoblasts was assessed by infecting the same cells with a retroviral vector encoding beta-galactosidase. This labeling method allowed both an assessment of cell number using a highly sensitive fluorogenic assay (Roederer et al., 1991) and localization of the cells within the tissue. In addition to labelled fibers formed by fusion between implanted myoblasts, a large proportion of labeled cells was found in large diameter fibers typical of the surrounding tissue, suggesting that implanted cells fused with the fibers of the host (Figure 2, bottom). Approximately 10^6 myoblasts were capable of producing and secreting 4 micrograms hGH/day in vitro and upon implantation in mice maintained a steady state serum level of 1 ng/ml. Taken together, these findings suggest that myoblasts show promise as vehicles for delivery to the circulation of a number of recombinant proteins.

CONCLUSIONS AND FUTURE PROSPECTS

The potential to isolate, characterize myoblasts in vitro, and reimplant them in vivo into the muscle of mouse and man will allow further examination of the complex processes underlying patterning, growth, and repair in normal and dystrophic muscle. The knowledge gained from these studies is not only of fundamental interest, but as described above, appears to have practical application to disease states. Although direct DNA injection has been proposed as an alternative to cell-mediated gene transfer, to date this approach is too inefficient to be useful in the delivery of recombinant proteins to the circulation (Lin et al., 1990; Acsadi et al., 1991, Kitsis et al., 1991). The challenge of the next decade will be to test and further refine myoblast-mediated gene delivery for the treatment of a range of inherited and acquired disease states.

REFERENCES

Acsadi G, Dickson G, Love DR, Jani A, Walsh FS, Gurusinghe A, Wolff JA and Davies KE (1991): Human dystrophin expression in mdx mice after intramuscular injection of DNA constructs. *Nature* 352: 815-818.

Anderson WF (1992): Human gene therapy. *Science* 256: 808-813.

Barr E and Leiden JM (1991): Systemic delivery of recombinant proteins by genetically modified myoblasts. *Science* 254: 1507-1509.

Blau HM and Hughes SM (1990): Cell lineage in vertebrate development. *Current Opinion in Cell Biology* 2: 981-985.

Blau HM, Kaplan I, Tao T-W and Kriss JP (1982): Thyroglobulin-independent, cell-mediated cytotoxicity of human eye muscle cells in tissue culture by lymphocytes of a patient with Graves' ophthalmopathy. *Life Sci.* 32: 45-53.

Blau HM and Webster C (1981): Isolation and characterization of human muscle cells. *Proc. Natl. Acad. Sci. USA* 78: 5623-5627.

Blau HM, Webster C, Chiu C-P, Guttman S and Chandler F (1983a): Differentiation properties of pure populations of human dystrophic muscle cells. *Exp. Cell Res.* 144: 495-503.

Blau HM, Webster C and Pavlath GK (1983b): Defective myoblasts identified in Duchenne muscular dystrophy. *Proc. Natl. Acad. Sci. USA* 80: 4856-4860.

Costa EM, Blau HM and Feldman D (1986): 1,25 dihydroxyvitamin D_3 receptors and hormonal responses in cloned human skeletal muscle cells. *Endocrinology* 119(5): 2214-2220.

Dhawan J, Pan LC, Pavlath GK, Travis MA, Lanctot AM, and Blau HM (1991): Systemic delivery of human growth hormone by injection of genetically engineered myoblasts. *Science* 254: 1509-1512.

Freidman T (1989): Progress toward human gene therapy. *Science* 244: 1275-1281.

Garver RI, Chytil A, Courtney M and Crystal RG (1987): Clonal gene therapy: Transplanted mouse fibroblast clones express human alpha-1 anti-trypsin in vivo. *Science* 237: 762-764.

Gunning P, Hardeman E, Wade R, Ponte P, Bains W, Blau HM and Kedes L (1987): Differential patterns of transcript accumulation during human myogenesis. *Mol. Cell. Biol.* 7: 4100-4114.

Gussoni E, Pavlath GK, Lanctot AM, Sharma K, Miller RG, Steinman L and Blau HM (1992): Normal dystrophin transcripts detected in DMD patients after myoblast transplantation. *Nature* 356: 435-437.

Ham RG, St. Clair JA, Webster C and Blau HM (1988): Improved media for normal human muscle satellite cells: Serum-free clonal growth and enhanced growth with low serum. *In Vitro Cell Dev. Biol.* 24: 833-844.
Hoffman EP, Morgan JE, Watkins SC and Partridge TA (1990): Somatic reversion/suppression of the mouse mdx phenotype in vivo. *J. Neurol. Sci.* 99: 9-25.
Hughes SM and Blau HM (1990): Migration of myoblasts across basal lamina during skeletal muscle development. *Nature* 345: 350-353.
Hughes SM, and Blau HM (1992): Muscle fiber pattern is independent of cell lineage in postnatal rodent development. *Cell* 68: 659-671.
Kaplan I, Blakely BT, Pavlath GK, Travis MA and Blau HM (1990): Steroids induce acetylcholine receptors on cultured human muscle: Implications for myasthenia gravis. *Proc. Natl. Acad. Sci. USA* 87: 8100-8104.
Kaplan IK and Blau HM (1986): Metabolic properties of human acetylcholine receptors can be characterized on cultured human muscle. *Exp. Cell Res.* 166: 379-390.
Karpati G, Pouliot Y, Zubrycka-Gaarn E, Carpenter S, Ray PN et al. (1989): Dystrophin is expressed in mdx skeletal muscle after normal myoblast implantation. *Am. J. Pathol.* 135: 27-34.
Kitsis RN, Buttrick PM, McNally ML, Kaplan ML and Leinwand LA (1991): Hormonal modulation of a gene injected into rat heart in vivo *Proc. Natl. Acad. Sci. USA* 88: 4138-4142.
Leiter EH, Beamer WG, Schultz LD, Barker JE and Lane PW (1987): Mouse models of genetic disease. *Birth Defects* 23: 221-257.
Lin H, Parmacek MS, Morle G, Bolling S and Leiden JM (1990): Expression of recombinant genes in myocardium in vivo after direct injection of DNA. *Circulation* 82: 2217-2221.
Mantegazza R, Hughes SM, Mitchell D, Travis MA, Blau HM and Steinman L (1991): Modulation of MHC Class II antigen expression in human myoblasts after treatment with IFN-γ. *Neurology* 41: 1128-1132.
Mauro A (1961): Satellite cells of skeletal muscle fibers. *J. Biophys. Biochem. Cytol.* 9: 493-495.
Miller AD (1990): Retrovirus packaging cells. *Hum. Gene Therapy* 1: 5-14.
Miller AD (1992): Human gene therapy comes of age. *Nature* 357: 455-460.

Miller SC, Ito H, Blau HM and Torti FM (1988): Tumor necrosis factor inhibits human myogenesis in vitro. *Mol. Cell. Biol.* 8: 2295-2301.

Morgan J, Barrandon Y, Green H and Mulligan RC (1987): Expression of an exogenous growth hormone gene by transplantable human epidermal cells. *Science* 237: 1476-1479.

Morgan RA, Looney D, Muenchau D, Wong-Staal F, Gallo R and Anderson WF (1990): Retroviral vectors expressing soluble CD4: A potential gene therapy for AIDS. *AIDS Res. and Human Retroviruses* 6: 183-191.

Palmer TD, Rossman G, Osborne WRA and Miller AD (1991): Genetically modified skin fibroblasts persist long after transplantation but gradually inactivate introduced genes. *Proc. Natl. Acad. Sci. USA* 88: 1330-1334.

Partridge TA, Grounds MD and Sloper JC (1978): Evidence of fusion between host and donor myoblasts in skeletal muscle grants. *Nature* 273: 306-308.

Partridge TA, Morgan JE, Coulton GR, Hoffman EP and Kunkel LM (1989): Conversion of mdx myofibers from dystrophin-negative to positive by injection of normal myoblasts. *Nature* 337: 176-179.

Roederer M, Fiering S and Herzenberg LA (1991): FACS-Gal: Flow cytometric analysis and sorting of cells expressing reporter gene constructs. *Methods* 2: 248-260.

Roman M, Axelrod JH, Dai Y, Naviaux RK, Friedmann T and Verma IM (1992): Circulating human or canine factor IX from retrovirally transduced primary myoblasts and established myoblast cell lines grafted into murine skeletal muscle. *Somat. Cell Mol. Genet.* 18: 247-258.

Scharfmann R, Axelrod JH, and Verma IM (1991): Long-term in vivo expression of retrovirus-mediated gene transfer in mouse fibroblast implants. *Proc. Natl. Acad. Sci. USA* 88: 4626-4629.

Selden RF, Skoskiewicz MJ, Hower KB, Russel PS, Goodman HM (1987): Implantation of genetically engineered fibroblasts into mice: Implication for gene therapy. *Science* 236: 714-717.

Shimizu M, Webster C, Morgan DO, Blau HM and Roth RA (1986) Insulin and insulin-like growth factor receptors and responses in cultured human muscle cells. *Amer. J. Physiol.* 251: E611-E615.

St Louis D and Verma IM (1988): An alternative approach to somatic cell gene therapy. *Proc. Natl. Sci. Acad. USA* 85: 3150-3154.

Teumer J, Lindahl A and Green H (1990): Human growth hormone in the blood of athymic mice grafted with cultures of hormone-secreting human keratinocytes. *FASEB J.* 4: 3245-3249.
Webster C and Blau HM (1990): Accelerated age related decline in replicative life-span of Duchenne muscular dystrophy myoblasts: Implications for cell and gene therapy. *Som. Cell & Mol. Genet.* 16: 557-565.
Webster C, Filippi G, Rinaldi A, Mastropaolo C, Tondi M, Siniscalco M and Blau HM (1986): The myoblast defect identified in Duchenne muscular dystrophy is not a primary expression of the DMD mutation: Clonal analysis of myoblasts from double heterozygotes for two X-linked loci: DMD and G6PD. *Human Genetics* 74: 74-80.
Webster C, Pavlath GK, Parks DR, Walsh FS and Blau HM (1988a): Isolation of human myoblasts with the fluorescence-activated cell sorter. *Exp. Cell Res.* 174: 252-265.
Webster C, Silberstein L, Hays AP and Blau HM (1988b): Fast muscle fibers are preferentially affected in Duchenne muscular dystrophy. *Cell* 52: 503-513.
Yao S-N and Kurachi K (1992): Expression of human factor IX in mice after injection of genetically modified myoblasts. *Proc. Natl. Acad. Sci. USA* 89: 3357-3361.

IMPLANTATION OF CULTURED SCHWANN CELLS TO FOSTER REPAIR IN INJURED MAMMALIAN SPINAL CORD

Mary Bartlett Bunge, Carlos L. Paino*, Cristina Fernandez-Valle

The Miami Project to Cure Paralysis, Departments of Cell Biology and Anatomy, and Neurological Surgery, University of Miami School of Medicine, Miami, FL USA

*Present Address: Department of Investigacion, Hospital Ramón y Cajal, Madrid, Spain.

INTRODUCTION

Until relatively recently, the prevailing view was that successful axonal regeneration could not occur in the central nervous system (CNS) of the adult mammal. Attempts at regeneration, termed "abortive regeneration," were known to occur at the site of injury but continued growth that was maintained was not observed. This contrasts with the peripheral nervous system where regeneration and functional recovery have long been known to occur. In fact, this knowledge led workers in the early 1900s to transplant the substratum for axon regrowth, peripheral nerve, into the CNS. The results were successful enough to cause Ramón y Cajal (1928) to conclude that CNS nerve cells possess the capacity to regrow axons if a suitable environment is provided. Despite subsequent studies over the next decades, convincing evidence of the growth potential of CNS neurons in the spinal cord was not published until 1980. Richardson et al (1980) clearly demonstrated that CNS axons regrew into pieces of peripheral nerve transplanted into a 5-10 mm gap in adult rat spinal cord.

Peripheral nerve contains axons that are surrounded either by Schwann cell cytoplasm (and thus are "ensheathed") or myelin sheaths formed by Schwann cells. These axons are termed unmyelinated or myelinated, respectively. Each Schwann cell produces a sleeve of basal lamina that is continuous along the succession of Schwann cells covering the entire length of the axon. When nerve injury occurs, the axons distal to the damage degenerate but the Schwann cells and basal lamina remain; it is within this tunnel of basal lamina filled with Schwann cells that the axons unfailingly regenerate. Thus, one of the candidates responsible for fostering axon regeneration in peripheral nerve is the Schwann cell.

Schwann cells express many attributes that suggest their suitability for transplantation into the spinal cord. Although Schwann cells are not normally residents of the spinal cord, they are able to function, i.e. form myelin, when introduced into this environment (e.g., Duncan et al, 1981). Schwann cells produce neurotrophins such as nerve growth factor (Bandtlow et al, 1987) and brain derived neurotrophic factor (Acheson et al, 1991) that support neuronal survival and outgrowth, and they synthesize and secrete extracellular matrix molecules (Bunge et al, 1986; Bunge, 1993) such as laminin which is known to support neuronal outgrowth. They express cell adhesion molecules and integrins, some of which have been shown to be involved in axonal growth on Schwann cells (Bixby et al, 1988). The presence of Schwann cells is required for axonal growth from some CNS neurons in vitro (reviewed in Bunge and Hopkins, 1991). Moreover, when Schwann cells have been transplanted into several areas of the brain, they have enabled axonal regeneration that otherwise would not have occurred (Kromer and Cornbrooks, 1985).

In addition to all these growth-promoting characteristics of Schwann cells, another reason to consider their candidacy for transplantation into the spinal cord is the long-term potential for autologous transplantation. A piece of sensory nerve could be removed from a spinal cord injured person to provide Schwann cells for expansion in number in culture. Once a sufficient number was obtained, they could be introduced into the spinal cord lesion if their efficacy to provide an appropriate bridge for regrowth had been established.

MATERIALS AND METHODS

Purified populations of Schwann cells were obtained from dorsal root ganglia of 16 day Sprague-Dawley rat embryos (Wood, 1976; see also Kleitman et al, 1991). They were added to purified cultures of dorsal root ganglion neurons (obtained from 15 d rat embryos) placed in two or three narrow parallel rows on the 5-10 μm thick polymerized collagen (type I) substratum of the dish. (This collagen, extracted from rat tail, is spread on the dish and polymerized with ammonium hydroxide vapor and allowed to dry before adding culture medium and cells.) Axons grew between the neuronal cell bodies and were ensheathed or myelinated by the added Schwann cells which also formed typical basal lamina. (The culture conditions employed are known to enable this degree of Schwann cell differentiation; Bunge et al, 1986.) After severing the nerve cell bodies from their axons, the axons quickly degenerated, leaving aligned Schwann cells and their extracellular matrix behind. The collagen substratum was loosened from the culture dish and rolled into a scroll (Kuhlengel et al, 1990) which was coated with plasma to prevent unrolling. This preparation provided a cylindrically shaped graft about 1 mm in diameter and 4-6 mm in length (Paino and Bunge, 1991). Alternatively, to compare the efficacy of dissociated Schwann cells that had not been cultured with neurons, purified populations of dissociated Schwann cells were placed on comparable polymerized and cultured collagen substrata and rolled into scrolls. Scrolls prepared similarly but devoid of Schwann cells were also transplanted (Paino and Bunge, 1991).

The scrolls were transplanted into thoracic level cavities (as in Fig. 1A in Paino and Bunge, 1991) in adult (Sprague-Dawley) female rat spinal cord created by a photochemical method (reviewed in Watson et al, 1993). Briefly, an argon-dye laser (Innova 70 System, Coherent Radiation, Inc., Palo Alto, CA) beam 100mW was focused transversely over the T8 lamina of the vertebral column immediately after a photosensitizing agent (rose bengal dye) had been injected intravenously. At 150 sec, 50% duty cycle irradiation produced a photothrombotic lesion that consistently led to degeneration within the dorsal half of the spinal cord. A fusiform shaped lesion cavity developed, reaching a rostral-caudal extent of 4-6 mm after 1 wk (Cameron et al, 1990). A pair of scrolls was inserted into each lesion cavity at 5 or 28 d after lesioning, through a small longitudinal incision made in the dura and cord following a laminectomy at T7 and T8. No immunosuppression was used.

Animals were perfused with fixative at 14, 28, 90, and 180 d after transplantation and processed for light microscopy (semi-thin plastic sections stained with toluidine blue or paraffin sections stained with Sevier-Munger silver stain) and for electron microscopy. This preparation included perfusion with buffered 4% paraformaldehyde combined with either 0.1% or 3.6% glutaraldehyde (Kuhlengel et al, 1990), obtaining 0.5-1.0 mm transverse slices at the scroll center and 2 mm rostral and caudal to it before further processing, further fixation in buffered glutaraldehyde followed by up to 24 hr in cold OsO_4, and embedding in Epon-Araldite (Electron Microscopy Sciences, Fort Washington, PA).

Our findings are based on 23 animals receiving Schwann cell-laden collagen scrolls, 3 control animals receiving acellular collagen scrolls, and 4 animals that were lesioned but did not receive scrolls. In 9 cases, a montage of an entire 1μm thick, plastic cross-section at the center of the two transplanted scrolls (at 800X magnification) was prepared in order to obtain the number of myelinated axons. A randomized method was also developed to obtain electron micrographs from 5 transplanted animals to calculate the ratio of unmyelinated to myelinated axons. The resolution of the electron microscope was required to detect the unmyelinated axons.

RESULTS

The grafted scrolls filled the lesion cavity (Fig. 1a). They closely apposed much of the cavity laterally and reached at least portions of the rostral and caudal ends of the cavity. Whereas occasional clumps of macrophages separated the graft and the host cord, the outermost layer of collagen of the scroll generally closely abutted the host tissue. The collagen could be seen for the duration of the longest observation period. In spite of the lack of immunosuppression, rejection or an adverse

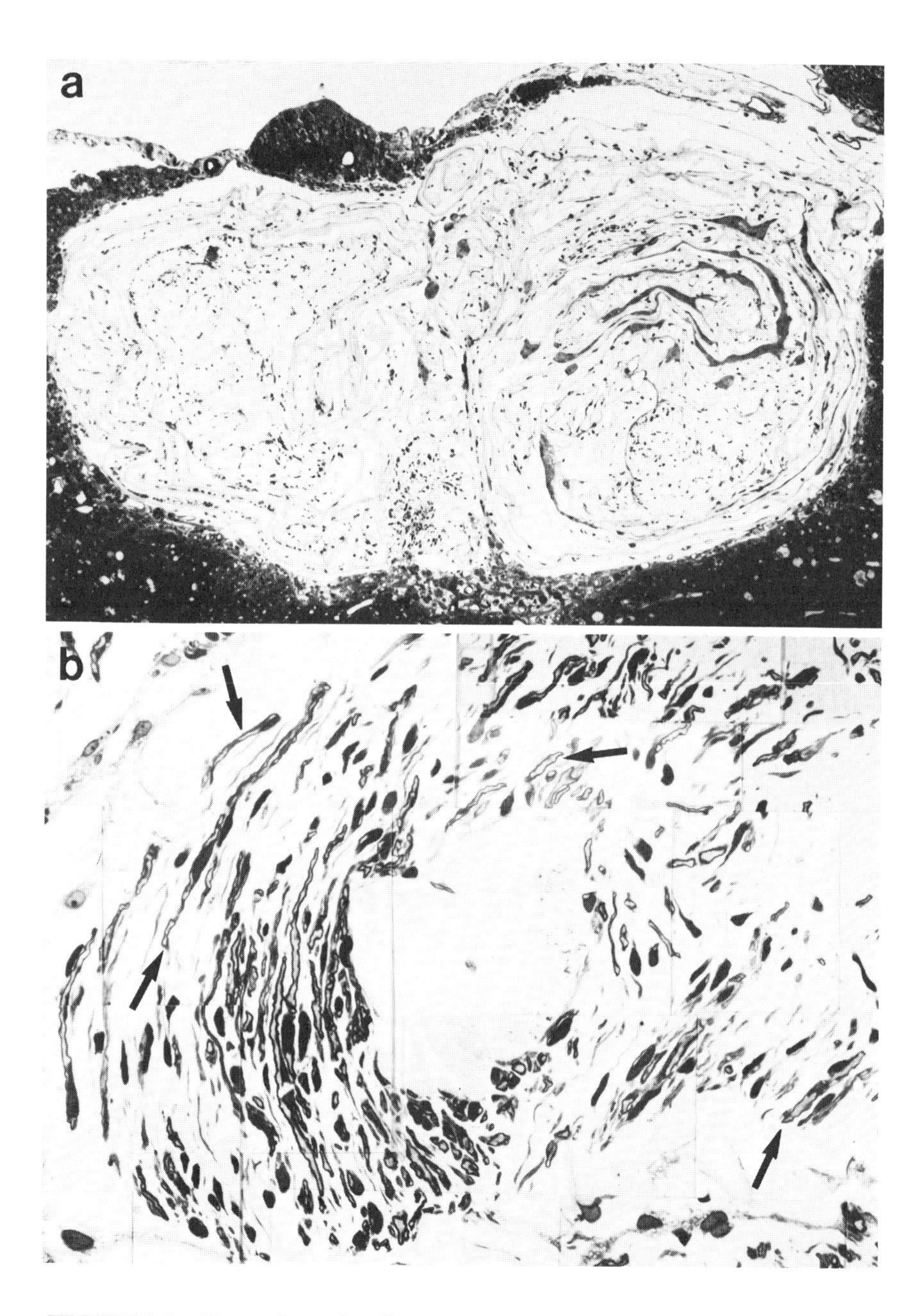

FIGURE 1. Transplanted collagen scrolls containing Schwann cells and their extracellular matrix. Panel a illustrates two scrolls that have filled the lesion cavity; they closely appose the surrounding host tissue. The dorsal surface of this cross-sectioned spinal cord is at the top of the panel; 80 X. Panel b depicts a portion of a montage prepared from another implanted scroll. Numerous myelinated axons (arrows) are present; 400 X. 28 d after transplantation.

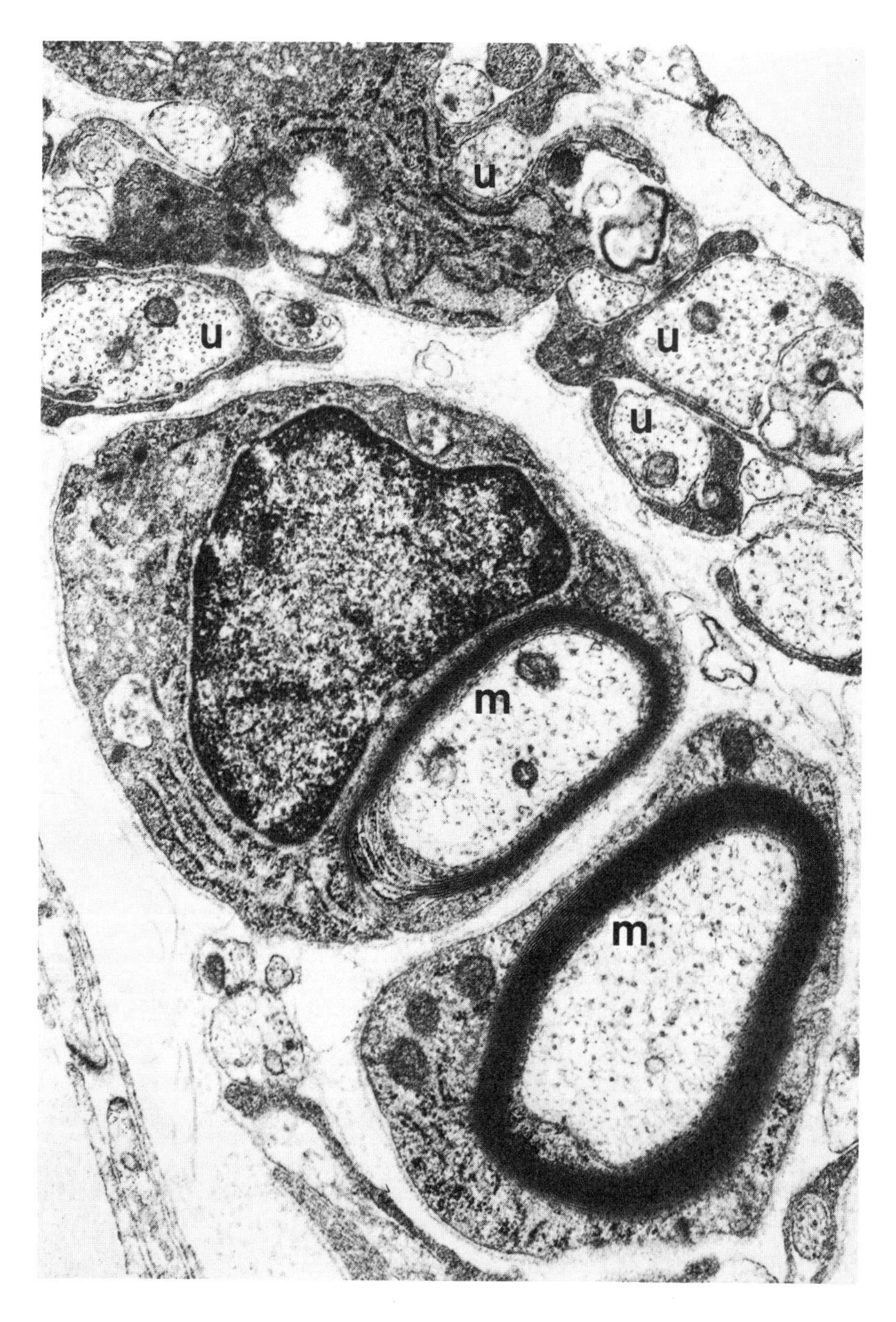

FIGURE 2. This electron micrograph of a transplanted collagen scroll containing Schwann cells and their extracellular matrix illustrates the presence of both myelinated (m) and unmyelinated (u) axons; the unmyelinated axons are surrounded by Schwann cell cytoplasm. 28 d after transplantation; 27,000 X.

immune reaction was not apparent up to 180 d after transplantation into this partially inbred rat strain. The host astrocyte reaction around the transplant perimeter appeared to be minimal, in contrast to increased astrogliosis at the lesion cavity border in the absence of a Schwann cell transplant (Salvatierra et al, 1990; Watson et al, 1993).

By 14 d after implantation, host axons had entered the outermost layers of the scroll and were just beginning to acquire myelin. Numerous cytoplasmic images seen at the light microscopic level were found in the electron microscope to be Schwann cell processes that had ensheathed but not myelinated axons. Myelinated axons were numerous by 28 d after grafting (Figs. 1b,2); 647 and 1,369 myelinated axons were found in two montages of a 1 μm thick cross-section of implanted scrolls that contained Schwann cells and their extracellular matrix.

The number of myelinated axons continued to increase. By 90 d after transplantation, 1,252 and 2,485 myelinated axons were found in two montages of cross-sectioned scrolls containing Schwann cells and their matrix. Myelin sheath thickness was increased as well. The ratio of unmyelinated to myelinated axons was 7.6 at this time point (Paino et al, 1991), indicating that axons numbered in the thousands at the midpoint of these transplanted scrolls. Growth of host axons into the implant of Schwann cells and collagen was, therefore, abundant. Silver stained preparations of longitudinally sectioned scrolls revealed long bundles of axons within the transplant. The content of axons at the later time periods resembled that at the 90 d post-grafting period.

Collagen alone, cultured and rolled as in the cell-containing collagen scrolls, was grafted into lesioned animals which were maintained for 28 d. Myelinated axons were not found inside the scroll (Fig. 1C and D in Paino and Bunge, 1991). Also, unmyelinated axons were not found in collagen implants that were carefully scrutinized by electron microscopy. Some fibroblast-like cells were observed to have entered the collagen only implant (Fig. 1D in Paino and Bunge, 1991). These findings indicated that Schwann cells must be present in the implant to elicit ingrowth of host axons.

DISCUSSION

The observations that we present here indicate that Schwann cell-containing scrolls engender abundant host axonal ingrowth when transplanted into lesion cavities created in adult rat spinal cord. The roll of polymerized collagen in which the Schwann cells were confined appeared to be a suitable carrier in that it was stable throughout the experimental period, it was well tolerated by the host, and its spiral layering may have aided in longitudinal axonal growth along the implant. On the other hand, even though axonal ingrowth was substantial, there were images suggesting that the collagen acted overall as a barrier. It is known that collagen polymerized as it was for this study is dense compared with other methods of preparation (Bunge et al, 1987). Bundles of axons coursed from the host into the scroll opening or through discontinuities in the collagen. Some longitudinal plastic sections of the ends of a transplant showed that the outermost layer of collagen was intact, with no axons penetrating it at that level. This plus the lack of rigidity of the scroll raised the possibility that the ends curved, thereby limiting axonal ingress or egress.

Clearly, the best method of presentation of the Schwann cells to the spinal cord remains to be determined. In addition to the method we have described here, Schwann cells have been injected as a suspension into a spinal cord lesion cavity (Martin et al, 1991). Substantial axon growth was observed. Many of the axons were considered to have grown in from the dorsal root ganglia because they stained for characteristic neuropeptides. We do not know the source of axons that grew into the collagen scrolls. About 50% of the axons that grew into the whole peripheral nerve transplants in spinal cord originated in the dorsal root ganglia rather than the spinal cord (Richardson et al, 1980). This illustrates another challenge we face in spinal cord transplantation: to determine the origin of the axons within Schwann cell transplants and, further, to discover what additional conditions may be introduced into Schwann cell transplants to foster axonal ingrowth from a variety of neuronal cell types, some of which we know do not respond to Schwann cells alone.

One new approach that we have initiated is to enclose Schwann cells within a PAN/PVC permselective guidance channel rather than a collagen roll (Xu et al, 1992). A length of spinal cord, three segments long, is removed and the rostral cut end is inserted into the open end of the channel; the other end was capped in initial experiments. The cable of aligned Schwann cells that forms within the channel is accessible to the host cord, and the channel provides a controlled environment for the continued trial of components that will elicit axonal regeneration. One month after grafting, a blood vessel-rich cable containing hundreds of myelinated axons was observed along the length of the channel (10 mm). Larger numbers of unmyelinated axons were present as well. Thus, when Schwann cells were presented to spinal cord in this new paradigm, they also stimulated ingrowth of axons. Schwann cell-laden channels have also been found to foster growth from retinal ganglion cells when the channels are used to bridge optic nerve (Guenard et al, 1991). The next few years will be an exciting time when new materials and varied modes of presentation will be utilized in the search for the most effective means to improve CNS regeneration after injury. Whereas we do not yet know what the most effective means will be, it is clear that in all methods of presentation employed thus far Schwann cells have exhibited a remarkable capacity to elicit CNS regenerative growth.

CONCLUSIONS

1. Cultures of Schwann cells can be prepared on supporting collagen and rolled into a scroll for transplantation into lesioned adult rat spinal cord. The implants fill the lesion cavity, are well tolerated by the host, and appear to minimize astrogliosis.

2. The implanted Schwann cells stimulate rapid and abundant growth of host axons into grafts, whereas acellular collagen grafts do not stimulate axon ingrowth.

3. The implanted Schwann cells ensheathe and begin to myelinate the host axons that grow into the grafts within 14 d after transplantation. Myelination continues over the next months and both the myelinated and more numerous ensheathed axons appear to be stable throughout the longest period examined, 180 d.

ACKNOWLEDGMENTS

We thank B. Watson, V. Holets, R. Prado, and A. Salvatierra for help with photochemical lesioning; M. Bates, J. Olivier, and J. Brown for excellent laboratory assistance, and particularly M. Bates for preparing the photographic montages and J. Olivier for surveying grafts electron microscopically to enable counting unmyelinated axons; and C. Rowlette for expert word processing. This work was supported by NIH grants NS28059 and NS 09923, The Miami Project, and a grant to CLP from the Spanish Ministry of Education.

REFERENCES

Acheson A, Barker PA, Alderson RF, Miller FD and Murphy RA (1991): Detection of brain-derived neurotrophic factor-like activity in fibroblasts and Schwann cells: Inhibition by antibodies to NGF. Neuron 7:265-275.

Bandtlow CE, Heumann R, Schwab ME and Thoenen H (1987): Cellular localization of nerve growth factor synthesis by *in situ* hybridization. EMBO J 6:891-899.

Bixby JL, Lilien J and Reichardt LF (1988): Identification of the major proteins that promote neuronal process outgrowth on Schwann cells in vitro. J Cell Biol 107:353-361.

Bunge MB (1993): Schwann cell regulation of extracellular matrix biosynthesis and assembly. In: *Peripheral Neuropathy* (3rd Edition), Dyck PJ, Thomas PK, Griffin JW, Low PA, Poduslo JF, eds. Philadelphia, WB Saunders. pp 299-316.

Bunge MB, Johnson MI, Ard MD and Kleitman N (1987): Factors influencing the growth of regenerating nerve fibers in culture. In: *Progress in Brain Research*, Seil FJ, Herbert E, Carlson BM, eds. Elsevier, 71:61-74.

Bunge RP and Hopkins JM (1990): The role of peripheral and central neuroglia in neural regeneration in vertebrates. IN: *Seminars in the Neurosciences. Neurobiology of Glia.* Jessen KR and Mirsky R, eds. Philadelphia, PA, WB Saunders Company, 2:509-518.

Bunge RP, Bunge MB and Eldridge CF (1986): Linkage between axonal ensheathment and basal lamina production by Schwann Cells. Ann Rev Neurosci 9:305-328.

Cameron T, Prado R, Watson BD, Gonzalez-Carvajal M and Holets VR (1990): Photochemically induced cystic lesion in the rat spinal cord. I. Behavioral and morphological analysis. Exp Neurol 109:214-223.

Duncan ID, Aguayo AJ, Bunge RP and Wood PM (1981): Transplantation of rat Schwann cells grown in tissue culture into the mouse spinal cord. J Neurol Sci 49:241-252.

Guénard V, Morrissey TK, Kleitman N, Bunge RP and Aebischer P (1991): Cultured syngeneic adult Schwann cells seeded in synthetic guidance channels enhance sciatic and optic nerve regeneration. Soc Neurosci Abstr 17:565.

Kleitman N, Wood PM and Bunge RP (1991): Tissue culture methods for the study of myelination. In: *Neuronal Cell Culture*, Banker GA and Goslin K, eds. Boston, MIT Press, pp. 337-377.

Kromer LF and Cornbrooks CJ (1985): Transplants of Schwann cell cultures promote axonal regeneration in the adult mammalian brain. Proc Nat Acad Sci USA 82:6330-6334.

Kuhlengel KR, Bunge MB, Bunge RP and Burton H (1990): Implantation of cultured sensory neurons and Schwann cells into lesioned neonatal rat spinal cord. II. Implant characteristics and examination of corticospinal tract growth. J Comp Neurol 293:74-91.

Martin D, Schoenen J, Delree P, Leprince P, Rogister B and Moonen G (1991): Grafts of syngeneic cultured, adult dorsal root ganglion-derived Schwann cells to the injured spinal cord of adult rats: Preliminary morphological studies. Neurosci Lett 124:44-48.

Paino CL and Bunge MB (1991): Induction of axon growth into Schwann cell implants grafted into lesioned adult rat spinal cord. Exp Neurol 114:254-257.

Paino CL and Bunge MB (1990): Axon growth into implants of Schwann cells placed in lesioned spinal cord. Soc Neurosci Abstr 16:1282.

Paino CL, Fernandez-Valle C and Bunge MB (1991): Axon growth into Schwann cell grafts placed in lesioned adult rat spinal cord. Soc Neurosci Abstr 17:236.

Ramón y Cajal S (1928): *Degeneration and Regeneration of the Nervous System*. May RM (Translator), New York, Hafner Publishing Co.

Richardson PM, McGuinness U and Aguayo AJ (1980): Axons from CNS neurons regenerate into PNS grafts. *Nature* 284:264-265.

Salvatierra AT, Holets VR and Bunge MB (1990): Characterization of rat spinal cord changes following photochemically-induced injury. Soc Neurosci Abstr 16:1282.

Watson BD, Holets VR, Prado R and Bunge MB (1993): Laser-driven photochemical induction of spinal cord injury in the rat: Methodology, histopathology, and applications. NeuroProtocols *(In Press)*.

Wood PM (1976): Separation of functional Schwann cells and neurons from normal peripheral tissue. Brain Res 115:361-375.

Xu XM, Guénard V, Kleitman N and Bunge MB (1992): Axonal growth into Schwann cell-seeded guidance channels grafted into transected adult rat spinal cord. Soc Neurosci Abstr 18:1479.

PROGENITOR CELLS IN EMBRYONIC AND POST-NATAL RAT LIVERS, THEIR GROWTH AND DIFFERENTIATION POTENTIAL

Normand Marceau, Claude Chamberland, Marie-Josée Blouin, Micheline Noël and Anne Loranger.

Laval University Cancer Center, Hotel-Dieu de Quebec Hospital, Quebec City, Quebec, G1R 2J6, Canada.

INTRODUCTION

The adult liver is a multicellular organ composed of hepatocytes, biliary epithelial cells, mesothelial cells and several types of non-epithelial cells. The organ exhibits many metabolic fonctions, the main ones being the production of bile and plasma proteins (Daoust and Brauer, 1958). Liver damage is frequent in humans, mostly from genetic alterations through inheritance or acute metabolic failures as a result of exposure to toxic chemicals (Bucher and McGowan, 1979). The most dramatic consequence of liver damage is liver failure (Martin and Feldmann, 1983) which requires organ transplantation, a procedure that has many obvious drawbacks (Keeffe, 1991). An alternative approach would be to transplant normal liver cells that have the potential of re-populating the organ and exerting the specialyzed functions of the mature hepatic tissue. This cell transfer approach has already been used with good success for other tissues, such as the hemopoietic tissue cell system(Thomas et al., 1992; Carlo-Stella et al., 1992), where rare stem cells have the capacity to grow massively and to differentiate along the various hemopoietic cell lineages.

Conclusive evidence for the presence of progenitor (stem) cells in adult liver is still missing (Sell, 1990). Hepatocytes and biliary epithelial cells constitute the two main populations of specialized epithelial cells of adult liver. Both cell types exhibit a very low turnover but are capable of rapid proliferation in response to tissue insults (Bucher and McGowan, 1979; Sirica et al., 1991). For example, following a two-thirds hepatectomy in adult rat liver, restoration of the lobular parenchyma is apparently due to the rapid growth of remaining hepatocytes (Bucher and McGowan, 1979). On these grounds, the need for a progenitor cell compartment is not obvious. Our laboratory is examining this important issue by using rat liver as a model system. Since the differentiation of hepatocytes is developmentally regulated and biliary ductal structures appear before birth, our working hypothesis is that progenitors of hepatocytes and biliary epithelial cells should represent the main population of the emerging rat embryonic liver (Germain et al., 1988; Marceau et al., 1992). We have therefore examined the developmental features of fetal rat liver epithelial cells *in situ* and in culture (Germain et al., 1988), and then looked for the presence of such cells in adult rat liver (Marceau et al., 1992; Blouin, 1991; Blouin et al., in preparation). Our approach was to determine the differential expression of several metabolic and structural markers at various developmental periods *in situ* and to assess *in vitro* the growth and differentiation potential of cells isolated at embryonic day 12 (E12). We then searched for the presence of E12-like cells in the regenerating liver of adult rats following a single injection of D-galactosamine (Blouin, 1991). Previous data on the growth capacity of post-natal rat hepatocytes in primary culture were also considered. Finally, we compared the differentiation status of liver tumor cells with that of E12 cells.

EXPERIMENTAL PROCEDURES

In situ analysis: Livers from E12 and post-natal rat livers were cut into small pieces, embedded in O.C.T. compound, frozen in liquid nitrogen, and then processed for cryostat sectioning and indirect immunofluorescence staining (Germain et al., 1988). We used antibodies directed against α-fetoprotein (AFP), albumin (Alb), cytokeratins (CKs) 8, 14, 18, 19, OV-6, vimentin, cell-surface components exposed on hepatocytes (HES_6) and biliary epithelial cells (BDS_7), and a new marker corresponding to CK 8 phosphorylated forms (BPC_5) (Marceau et al., 1992; Blouin, 1991; Blouin et al., in preparation).

E12-cell culture: Cells were isolated by enzymatic digestion with a mixture of collagenase, dispase I and hyaluronidase, and seeded on fibronectin-coated glass coverslips distributed in 24-well culture plates in the presence of α-minimal essential medium, supplemented with 10% fetal calf serum (Germain et al., 1988). At day 1 post-seeding, the medium was replaced by a serum-free medium containing insulin, EGF, and dexamethasone in the presence or absence of sodium butyrate, dimethylsulfoxide, TGF-β, IGF-II or BRL-3A (Blouin, 1991). The differential expression of the various markers was then examined at days 1, 3 and 6 post-seeding.

RESULTS AND DISCUSSION

Embryonic rat liver.

In situ: E12 cells expressed AFP, Alb, CK8, CK18 and BPC_5. Beyond E16, biliary epithelial cells, which were organized as a few ductal islands, expressed BDS_7, CK 8, CK 18, CK 19 but not BPC_5 nor HES_6. Hepatocytic cells, which had typical bile canaliculi, expressed only HES_6, AFP, Alb, CK8 and CK 18 (Germain et al., 1988; Blouin, 1991). Cell typing analysis performed at various developmental stages led to the following observations: CK 8 and CK 18 were expressed in all epithelial cell types of the liver at all stages. CK 19 was always restricted to biliary ductal structures (Germain et al., 1988), whereas CK14 was found in some BDS_7-positive cells of the portal area in post-natal liver and also in cells of the Glisson's capsule (Blouin et al., 1992). Interestingly, CK14-positive cells of the portal area were vimentin negative, whereas those of the Glisson's capsule were positive (Blouin et al., 1992).

In culture: At day 1 post-seeding, the predominant epithelial cells expressed CK 8 and CK 18, but rarely CK 19 (Germain et al., 1988). They did not express vimentin but contained AFP and Alb. None of the cells were HES_6 positive, whereas a few were BDS_7 positive. At day 6, the use of a BRL-3A conditioned medium led to the emergence of a hepatocytic phenotype characterized by the appearance of HES_6, the maintenance of CK 8, CK 18, AFP and Alb and the decrease of BPC_5 (Table 1, Blouin, 1991). EM analysis revealed cells exhibiting typical ultrastructural features of small diploid hepatocytes (Germain et al., 1988). The most profound changes occurred when sodium butyrate was added (Germain et al., 1988; Marceau, 1990; Marceau et al., 1992; Blouin, 1991; Blouin et al., in preparation). Indeed, a large

MARKERS	PROGENITORs[b]	HEPATOCYTEs[c]	BILIARY ECs[c]
CK7	-	-	+
CK8	+	+	+
CK14	-	-	-
CKX	+	-	-
CK18	+	+	+
CK19	-	-	+
BPC_5	+	-	-
Alb	+	+	-
AFP	+	+	-
GGT	+	-	+
BDS_7	-	-	+
HES_6	-	+	-

[a]Modified from reference 54. According to Moll's et al human cytokeratin catalog[60], CK55 is equivalent to CK8 and CK 39 to CK19. The human equivalent of CK52, as reported in reference 54, is ill defined. While we proposed before that it might be equivalent to CK14[5], our recent data clearly establish that the CK detected in progenitors is a new CK, designated for now as CKX.[74] Alb denotes albumin.

[b]Progenitors correspond to epithelial cells present in the embryonic liver at the E10.5 to E16 period and specifically expressing BPC

[c]Hepatocytes and biliary ECs appearing after E15.

TABLE 1. In situ phenotypic properties of progenitors, hepatocytes and biliary ductal cells. Reprinted with permission of CRC Press Boca Raton, Fl. from A. Sirica (1992), chapter 7

proportion of the cells became BDS_7 and OV-6 positive, without losing their CKs, AFP or Alb. These cells displayed typical ultrastructural features of biliary ductal cells (Table 1, Germain et al., 1988; Blouin, 1991).

The capacity of E12 cells to proliferate in culture was assessed under conditions that provided various mixtures of promoting factors which in other epithelial cell systems exhibit very potent growth stimulation activities (Blouin, 1991). The data we accumulated so far showed that whatever the combinations of factors we used, E12 cells exhibited a relatively low capacity to proliferate, as analyzed by ^{3}H-dThd or BrdU incorporation (Blouin, 1991, Chamberland et al., unpublished data). The typical growth response curve was characterized by a slower increase toward a maximum at day 3 post-seeding (Chamberland et al., unpublished data). However, the transition of E12 cells to hepatocytic or biliary epithelial cells can occur in the absence of cell growth activity (Germain et al., 1988). We concluded, therefore, that E12 cells constituted a population of bipotential progenitor cells for hepatocytes and biliary ductal cells (Figure 1, Germain et al., 1988; Marceau, 1990; Marceau et al., 1992).

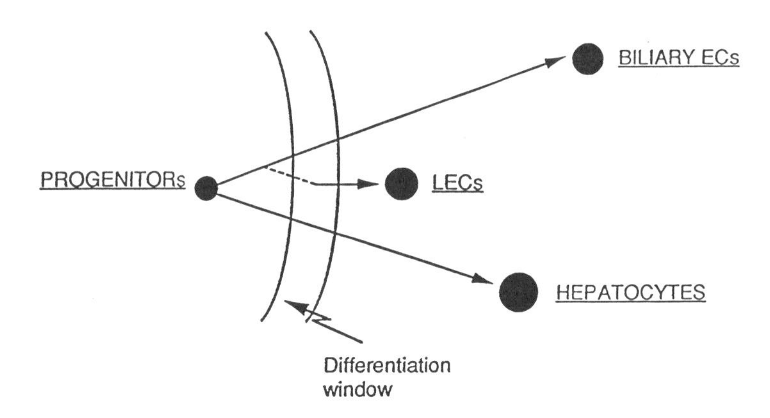

FIGURE 1. Transformation susceptibility of differentiating rat liver cells. Bipotential progenitor cells differentiating along hepatocytic or biliary ductal cell lineage can be maintained at a "differentiation window" that allows them to undergo transdifferentiation. At that stage, the cells are preferentially susceptible to transformation. Reprinted with permission of CRC Press Boca Raton, Fl. from A. Sirica (1992), chapter 7

Another observation of particular interest was made with regard to the cellular composition of the culture. Indeed, after a few days in primary culture, particularly in the presence of serum, we observed the appearance of non-parenchymal liver epithelial cells (LECs), which emerged as rapidly growing colonies of cells that lacked AFP, Alb, BDS_7, and HES_6 but still expressed CK8 and CK14 (Blouin et al., 1992).

Post-natal rat liver

in situ: At post-natal day 4 in the rat, the liver is undergoing its final three dimensional organization, which means that hepatocytes (CK8, CK18, AFP, Alb and HES_6 positive cells) are organized as single plates and biliary epithelial (CK7, CK8, CK18, CK19 and BDS_7 positive) cells form typical ducts and ductules (Table 1, Vassy et al., 1988). Already at that period, the relative number of BPC_5-positive cells had decreased to a low value (Blouin, 1991), which means that the E12-like cells were rapidly lost after birth.

At post-natal day 14 , the hepatocytes (CK8 and CK18 positive cells) have completed their maturation, as monitored by the loss of AFP (Baribault et al., 1985). Beyond that age, they undergo polyploidization, a sort of terminal differentiation that is linked to a higher production of Alb (Deschênes et al., 1981).

In culture: We have shown before that hepatocytes from post-natal rats seeded on fibronectin-coated coverslips in medium of widely different compositions, supplemented with various combinations of growth promoting factors exhibited a very low capacity to proliferate (Baribault et al., 1985). Sirica and his co-worker have reported that biliary ductal cells from bile duct-ligated rats plated in the presence of serum, can undergo several divisions but loose some of their markers, such as CK19 (Mathis et al., 1988). LECs can also be obtained as rapidly growing colonies in primary culture of post-natal rat hepatocytes (Marceau et al., 1986). These cells exhibited the phenotypic features of those that emerged in E12-cell cultures (Blouin et al., 1992).

Isom and her co-workers have subjected enriched preparations of adult rat hepatocytes to SV40 large T transfection under culture conditions that do not normally allow the cells to grow and were able to obtain slow growing colonies (foci) of highly Alb-producing cells (Woodworth et al., 1986). These cells could be passaged and were found to express several enzyme activities and to produce many plasma proteins. A cell typing analysis using the battery of markers described above led to the following findings: SV40-transformed cells contained CK8 and CK18, as it is the case for mature hepatocytes, but they still expressed BPC_5 in the absence of HES_6 and the presence of BDS_7 at a low level (Marceau et al., 1992). We concluded that the SV40-transformed cells corresponded to differentiating bipotential progenitor cells arrested at a few steps along the hepatocytic cell lineage (Figure 1, Marceau et al., 1992).

Liver cancer

Many studies have been performed on the definition of the cellular pathways that lead to the emergence of liver cancers with the aim of identifying the target cells of carcinogens. The two main types of liver cancers are cholangiocarcinomas and hepatocellular carcinomas, and cells derived from both tumors can be established as cell lines (Marceau et al., 1992). Our cell typing analysis using the various markers described above revealed that the cholangiocarcinoma tissue and its derived cell lines both express CK7, CK8, CK18, CK19, and BDS_7, which are typical markers of biliary epithelial cells (Marceau et al., 1992). However, a more complex picture was obtained for hepatocellular carcinoma cells, the main findings being that while all cell lines expressed CK8 and CK18, HES_6 and vimentin were rarely expressed and BDS_7 was frequently present (Marceau et al., 1992). All of the cell lines expressed BPC_5, the antigenic marker of E12 cells. We concluded that cholangiocarcinoma and hepatocellular carcinoma cells can be aligned with bipotential progenitors which are committed to the hepatocytic or biliary epithelial cell lineage, but which remain at early differentiation stages that we defined as a "differentiation window" (Figure 1; Marceau et al., 1992).

CONCLUSION

The data we accumulated so far indicate that the emerging rat liver contains bipotential progenitor cells that can differentiate along either the hepatocytic or biliary epithelial cell lineage; the latter gives rise to a LEC sublineage. However, the number of these bipotential progenitor cells decreases rapidly after birth and so far, no E12-like cells were detected in regenerating rat liver (Marceau et al., 1992; Blouin, 1991). These data have fundamental implications on the possibility of obtaining an enriched population of progenitor cells from adult liver. Our current interest focuses on the characterization of a sub-population from the LEC compartment, that exhibits a high proliferating activity and variable hepatic features.

ACKNOWLEDGEMENTS

The research from our laboratory reviewed here was supported by grants from the Medical Research Council of Canada and the National Cancer Institute of Canada.

REFERENCES

Baribault H. Leroux-Nicollet I and Marceau N. (1985). Differential responsiveness of cultured suckling and adult rat hepatocytes to growth-promoting factors: entry into S phase and mitosis. *J. Cell. Physiol.* 122:105.

Blouin MJ. (1991). Définition des voies de différenciation des cellules épithéliales hépatiques chez le rat. *Ph.D. thesis.* Laval University. Quebec.

Blouin MJ, Germain L, Noël M, Dunsford H and Marceau N. (*in preparation*). Common antigenic features between rat liver bipotential progenitors and tumor cells.

Blouin R, Blouin MJ, Royal I, Grenier A, Roop DR and Marceau N. (1992). Selective cytokeratin 14 gene expression in rat liver nonparenchymal epithelial cells. *Differentiation.* In press.

Bucher NLR and McGowan JA. (1979). Regulatory mechanisms in liver regeneration. Wright R, Alberti GGMM, Karsan S, Millward-Sadler H eds. In: *Liver and biliary disease:* a pathophysiological approach. Philadelphia:WB Saunders Co.: 210.

Carlo-Stella C, Mangoni L, Almici C and Garau D. (1992). Differential sensitivity of adherent CFU-blast, CFU-mix, BFU-E and CFU-GM to mafosfamide: implications for adjusted dose purging in autologous bone marrow transplantation. *Exp. Hematol.* 20:328.

Chamberland C, Blouin MJ, Loranger A and Marceau N. Unpublished data.

Daoust R and Brauer RE. ed. (1958). *Liver function.* American Institute Biological Science. Waverly Press: 3.

Deschênes J, Valet JP and Marceau N. (1981). The relationship between cell volume, ploidy, and functional activity in differentiating hepatocytes. *Cell Biophys.* 3:321.

Germain L, Blouin MJ and Marceau N. (1988). Biliary epithelial and hepatocytic cell lineage relationships in embryonic rat liver as determined by the differential expression of cytokeratins, α-fetoprotein, albumin, and cell surface-exposed components. *Cancer Res.* 48:4909.

Keeffe EB. (1991). Liver transplantation--challenges for the future. *West J. Med.* 155:541.

Marceau N, Germain L, Goyette R, Noël M and Gourdeau H. (1986). Cell of origin distinct cultured rat liver epithelial cells, as typed by cytokeratin and surface component selective expression. *Biochem. Cell Biol.* 64:788.

Marceau N. (1990). Cell lineages and differentiation programs in epidermal, urothelial and hepatic tissues and their neoplasms. *Lab. Invest.* 63:4.

Marceau N, Blouin MJ, Noël M, Török N and Loranger A. (1992). *The role of bipotential progenitor cells in liver ontogenesis and neoplasia.* In: The role of cell types in heaptocarcinogenesis. A. Sirica ed. CRC press Inc. Boca Raton. In press.

Martin E and Feldmann G. (1983). In: *Embryologie-Anatomie. Méthodes d'étude macroscopique. Malpositions et malformations.* Masson ed. Histopathologie du foie et des voies biliaires. New-York: 1.

Mathis GA, Walls SA and Sirica AE. (1988). Biochemical characteristics of hyperplastic rat bile ductular epithelial cells cultured "on top" and "inside" different extracellular matrix substitutes. *Cancer Res.* 48:6145.

Sell S. (1990). Is there a liver stem cell? *Cancer Res.* 50:3811.

Sirica A, Elmore LW and Sano N. (1991). Characterization of rat hyperplastic bile ductular epithelial cells in culture and *in vivo. Digestive diseases and sciences.* 36:494.

Thomas TE, Abraham SJ, Phillips GL and Lansdorp PM. (1992). A simple procedure for large-scale density separation of bone marrow cells for transplantation. *Transplantation.* 53:1163

Vassy J, Kraemer M, Chalumeau MT and Foucrier J. (1988). Development of the fetal rat liver: ultrastructural and stereological study of hepatocytes. *Cell differentiation.* 24:9.

Woodworth C, Secott T and Isom HC. (1986). Transformation of rat hepatocytes by transfection with simian virus 40 DNA to yield proliferating differential cells. *Cancer Res.* 46:4018.

Section IV

Co-Cultures and Other *In Vitro* Systems for Promoting Differentiation and Tissue Formation

EXTRACELLULAR MATRIX, CELLULAR MECHANICS AND TISSUE ENGINEERING

Donald Ingber
Departments of Pathology and Surgery,
Children's Hospital, Harvard Medical School

INTRODUCTION

One of the major recurring themes in this symposium on Tissue Engineering is that to design effective artificial tissues, we must first understand the critical chemical and structural determinants that control tissue development. Current tissue engineering approaches commonly use cell attachment scaffolds that are complex composites of naturally occuring extracellular matrix (ECM) molecules (e.g., collagens, glycosaminoglycans). Unfortunately, these "artificial ECMs" are restricted from an engineering standpoint: they exhibit a limited range of structural and chemical properties and are not easily chemically modified. Also, their large-scale fabrication can be limited by "batch to batch" variability during purification of the individual ECM molecules. An alternative approach for cell transplantation is to develop a completely synthetic attachment foundation that can support a high degree of cell function and yet be highly biocompatible (Vacanti et al., 1988; Cima et al., 1991). To accomplish this objective, recent advances in ECM biology must be merged with new developments in bioengineering and polymer chemistry.

We have focused our efforts on developing a simplified in vitro model to analyze how ECM regulates histogenesis. Our objective was to reduce the problem of tissue development down to "one molecule of problem", that is, to simplify the system as much as possible without losing critical structures, regulatory molecules, or a physiologically relevant response. Specifically, we set out to determine directly whether or not altering ECM type, orientation or mechanics was critical for control of cell sensitivity to soluble mitogens. With this simplified system, we also could separate effects on cell growth and function induced by ECM from those caused by cell-cell interactions. This experimental design is in direct contrast to approaches which attempt to recreate the complexity of the normal tissue microenvironment within a culture dish (Stoker et al., 1990). This latter method is often very effective at maintaining specialized cell functions. However, the presence of multiple growth factors, ECM components, and unknown contaminants (e.g., when complex ECM extracts are used) greatly complicates analysis of the molecular basis of tissue regulation.

Thus, using our more simplified approach, we set out to identify the minimal molecular and structural determinants that are necessary to sustain cell growth and differentiation. Once identified, these critical determinants could be used as design criteria for development of future artificial ECMs for specialized tissue engineering applications. The simplified model systems themselves also might be useful as in vitro drug screening assays or toxicology testing kits. Our results using this experimental approach and their implications for tissue engineering are described below.

CHEMICAL VERSUS MECHANICAL DETERMINANTS

Over the past decade, great advances have been made in terms of increasing our knowledge of the molecules that control tissue formation. Soluble growth factors and insoluble ECM molecules have been isolated. Transmembrane signaling pathways have been identified. Genes that control growth, differentiation, and even pattern formation have been cloned. Nevertheless, the mechanism of tissue genesis, the process by which three dimensional form is generated, remains unclear.

However, what is becoming increasingly clear is that while chemicals mediate this process, mechanical forces also play a major regulatory role. For example, it has been known for over 75 years that bone matrix is not deposited randomly. Rather it appears in specific patterns that correspond precisely to engineering lines of tension and compression for any structure of that size and shape with similar load-bearing characteristics (Koch, 1917). We now know that bone is a dynamic structure and that remodeling occurs at the cell level through the action of osteoblasts and osteoclasts. Thus, somehow these cells must be able to recognize and respond to changes in mechanical forces such that they remodel the bone ECM until local stress is minimized. Importantly, stress-induced remodeling of ECM and tissue form appears to be a general property of all living tissues (Ingber and Jamieson, 1985; Ingber, 1991).

Thus, to explain how growth and differentiation are coordinated during tissue development, we must first understand how both types of signals, chemical and mechanical, are integrated inside the cell. Luckily, this process of signal integration can be studied in vitro. For example, it has been known for many years that cells differ in their ability to grow and differentiate depending on both the chemistry and mechanics of their ECM substratum (Fig. 1). Epithelial cells commonly express more differentiated functions when grown on immobilized ECM molecules (e.g., collagen, fibronectin, laminin) compared to tissue culture plastic alone (Michalopoulos and Pitot, 1975; Ingber et al., 1987; Bissell et al., 1987;

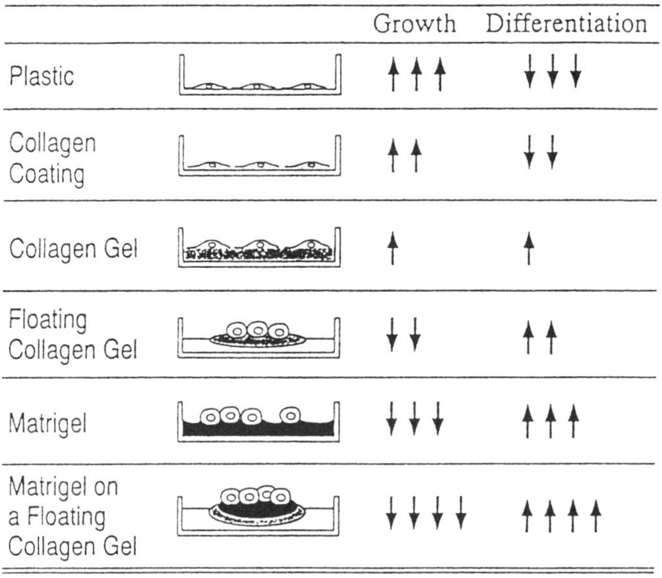

		Growth	Differentiation
Plastic		↑↑↑	↓↓↓
Collagen Coating		↑↑	↓↓
Collagen Gel		↑	↑
Floating Collagen Gel		↓↓	↑↑
Matrigel		↓↓↓	↑↑↑
Matrigel on a Floating Collagen Gel		↓↓↓↓	↑↑↑↑

FIGURE 1. Importance of ECM Chemistry and Mechanics. See text for description.

Sawada et al., 1987; Ingber and Folkman, 1989a; Reid, 1990). These cells function even better when maintained on specialized basement membrane substrata, such as Matrigel (Li et al., 1987; Bissell et al., 1987; Ben Ze'ev et al., 1988). Furthermore, ECM molecules produce different effects when presented in different structural configurations: ECM components that stimulate proliferation when coated on rigid plastic dishes suppress growth and promote differentiation when presented as a malleable gel (Michalopoulos and Pitot, 1975; Bissell et al., 1987; Li et al., 1987; Ben Ze'ev et al., 1988; Opas, 1989; Mochitate et al., 1991). These effects on differentiation can be further enhanced by releasing attached ECM gels from their attachments to the dish and allowing them to float free in the culture medium (Emerman and Pitelka, 1977; Li et al., 1987). Under these conditions, the gels physically retract as a result of the action of the adherent cells which exert tension on their ECM adhesion sites (Emerman and Pitelka, 1977; Bell et al., 1979; Harris, 1982; Ingber and Jamieson, 1985).

It is important to note that Matrigel, which is a complex of different basement membrane molecules that commonly supports cell differentiation (Sawada et al., 1987; Stoker et al., 1990), is also a malleable gel. However, even the differentiation-inducing effects of Matrigel are further enhanced when it is allowed to float free and permitted to retract completely (Li et al., 1987). The converse is also true: cross-linking Matrigel prevents both gel retraction and differentiation (Li et al., 1987; Opas, 1989).

A MECHANOCHEMICAL SWITCHING MECHANISM

If we could understand how changes in the chemistry and mechanics of the substratum switch cells between growth and differentiation, we would be in a much better position to design and construct artificial ECMs for tissue engineering applications. Unfortunately, at least four different explanations have been raised to explain why cell function is altered when ECM gels physically retract: 1) specific ECM binding sites are exposed (*chemical regulation*), 2) cell-cell interactions are enhanced (*junctional control*), 3) the orientation of the ECM changes relative to the cell (*geometric regulation*) and 4) the cellular force balance is altered such that cell shape changes result (*mechanical regulation*).

This last possibility, that the cellular force balance is a critical regulatory determinant, is based on the observation that cytoskeletal microfilaments that stretch between the cell's basal focal adhesions (i.e., ECM attachment sites) are contractile (Kreis and Birchmeier, 1980). For this reason, both retraction of ECM gels and associated changes in cell shape may be due to cells applying mechanical tension to their ECM adhesions (Ingber and Jamieson, 1985; Ingber, 1991; Sims et al., 1992). This is exactly what is observed within living cells cultured on thin silicone elastic substrata that are highly elastic (Harris, 1982) or, as I described above, when cells are plated on malleable ECM gels that also induce differentiation and shut off growth (Emerman and Pitelka, 1977; Ben Ze'ev et al., 1988; Mochitate et al., 1991). The point here is that much of the information conveyed by changing ECM mechanics may be mediated by changes in cell shape or cytoskeletal tension (Ingber and Jamieson, 1985; Ingber and Folkman, 1989a-c).

Another way to alter the cellular force balance and change cell shape is to vary the density of the cell's ECM attachment points and thus, the number of sites that can physically resist cell tractional forces. We first did this with cultured endothelial cells (Ingber and Folkman, 1989a; Ingber, 1990) by pre-coating bacteriological plastic with varying amounts of a single type of purified ECM molecule, such as fibronectin (FN). Cells can not attach to these dishes in the absence of adsorbed proteins, if serum is excluded from the medium. Thus, in all of these experiments, cells were cultured in chemically-defined medium supplemented with a constant, saturating concentration of a single type of recombinant endothelial mitogen, FGF.

These studies revealed that both DNA synthesis and cell proliferation increase in proportion as FN coating densities are raised and cell spreading is promoted (Ingber, 1990). Furthermore, low to moderate FN coating concentrations that promote cell retraction and rounding also induce differentiation (capillary tube formation) in high density cultures (Ingber and Folkman, 1989a). Importantly, this type of control over growth and function was not limited to bovine capillary cells. Dave Mooney and Linda Hansen in my group recently demonstrated similar shape-dependent switching between growth and differentiation in primary rat hepatocytes (Mooney et al., 1992a). These cells maintained high levels of differentiated functions (e.g., secretion of albumin, fibrogen, and transferrin) when grown on a low ECM density that does not permit cell spreading, even in the absence of significant cell-cell contacts. In contrast, high ECM densities that promoted extensive cell spreading induced cell division and suppressed differentiation. Furthermore, similar control was obtained regardless of the type of ECM molecule used for cell attachment; dishes coated with different densities of FN, laminin, and types I or IV collagens produced nearly identical effects on both growth and expression of liver-specific functions. In a separate study, we found that dishes coated with a synthetic RGD-peptide (Peptite 2000; Telios Labs) that promoted attachment but not cell shape changes also maintained high levels of differentiated function and induced a virtual state of quiescence within cultured hepatocytes (Mooney et al., 1992b). Again, this control over growth and differentiation was exerted solely by varying cell-ECM contacts within cells that were cultured in medium containing a constant and saturating concentration of soluble growth factors (EGF and insulin).

What does this all mean in terms of the original question of how switching between growth and differentiation is controlled? Clearly, the information that is responsible for modulating cell sensitivity to soluble mitogens and controlling cell growth and function is not localized within one specific ECM molecule or an individual binding site. Cell-cell interactions are also not critical for switching between these two developmental programs, although they may be important later in the specialization process (e.g., during capillary tube closure). The three dimensional geometry of the ECM and orientation of the ECM relative to the cell are not major factors since rigid plastic dishes coated with a low ECM density that permit cell rounding are as effective at supporting hepatocyte function as Matrigel (Ben Ze'ev et al., 1988) or type I collagen overlays (Dunn et al., 1989).

So in terms of switching between growth and differentiation programs, it seems that mechanical interactions between cell surface receptors and immobilized ECM molecules and associated changes in cell shape may be the most critical regulatory determinants. Thus, these experiments raise the possibility that someday in the future we may be able to engineer entirely synthetic ECM scaffoldings with defined chemical and structural properties. However, to do this most effectively, we will need to understand the molecular basis of this mechanochemical switching mechanism.

CHEMICAL AND MECHANICAL SIGNAL INTEGRATION

How does physically pulling on the cell's ECM attachment sites and altering cell shape influence cell function? The simplest explanation is that cell shape changes result in alterations in growth factor binding. For example, cells could express increased numbers of receptors on their surface as they extend. However, this does not seem to be the case. It was shown many years ago that cells do not increase their membrane surface area when they spread, rather they smooth microvillar projections and unfold this membrane reservoir (Erickson and Trinkaus, 1976). Furthermore, in capillary endothelial cells, we found that FN does not alter FGF binding or transmembrane signaling across growth factor receptors (Ingber et al., 1990). Instead, FN controls growth directly by altering the set-point of chemical signaling pathways inside the cell. Working with Martin Schwartz (Scripps Institute) and Claude LeChene (Harvard Medical School), we found that binding of FN to transmembrane integrin receptors activates the Na^+/H^+ antiporter on the cell surface (Ingber et al., 1990; Schwartz et al., 1991a,b). The same chemical signaling system is also activated by virtually all growth factors within minutes after they bind their own receptors. When activated, intracellular protons are exchanged for extracellular Na^+ and thus, an increase in intracellular pH results.

Integrins are members of a superfamily of transmembrane dimeric proteins that function as cell surface receptors for a wide variety of different ECM proteins (Albelda and Buck, 1990). Essentially, what we found is that binding of FN to specific integrins (e.g., ß1,α5) is responsible for activation of the antiporter and that this effect is required for growth (Ingber et al., 1990; Schwartz et al., 1991a,b). Occupancy of integrins did not activate this pathway; integrin clustering was required. Importantly, this mechanism of integrin activation differs from that observed with growth factor receptors. Growth factor-activated signaling mechanisms are usually turned on in response to receptor internalization. In contrast, signal transmission rapidly shuts off following internalization of integrin receptors. In other words, the ECM ligand has to both cluster integrins and physically resist receptor internalization (i.e., cytoskeletal tension) in order to produce a sustained growth signal. This may partially explain why ECM molecules have to be insoluble in order to promote cell proliferation.

More recently, we have found that binding of FN to integrins also controls inositol lipid metabolism (McNamee et al., 1993), activates tyrosine kinases (Plopper et al., 1991), and induces expression of early growth response genes (Hansen et al., 1991) in the absence of growth factors or a change in cell shape. The point of all this is that ECM molecules hook up to the same chemical signaling pathways and thus, the same maze of complexity that soluble growth factors utilize (Schwartz, 1992). ECM and growth factors work hand-in-hand to control these signaling events and thus, both are required for optimal cell proliferation.

However, while activation of these signaling pathways may be necessary for growth, I do not want to give the impression that it is sufficient; because it is not. In endothelial cells, the antiporter and other growth-related signaling events can be activated in round cells by allowing them to bind FN-coated microbeads that induce integrin clustering without producing global cell shape changes (Schwartz et al., 1991b). However, these cells do not enter S phase or grow (Ingber, 1990). Similarly, hepatocytes attach but neither spread nor grow on dishes coated with high densities of RGD-containing peptide, a substratum that should induce integrin clustering locally (Mooney et al., 1992b).

In summary, we have found that integrin-clustering can activate intracellular signaling events and promote cell cycle progression through the early part of the G_1 phase of the cell cycle. It is also sufficient to turn on differentiation-specific genes (Mooney et al., 1992a). However, for cells to pass out of this "default" differentiation pathway and to enter S phase, large-scale changes of cell shape are also required. The importance of these findings is that they suggest that ECM molecules also may convey critical regulatory information by activating what we call a "mechanical signaling" pathway inside the cell.

The concept of a mechanical signaling system emerged out of studies with three dimensional cell models that are built out of sticks and elastic string and are constructed according to the rules of an architectural system that depends on tensional integrity (tensegrity) (Ingber and Jamieson, 1985; Ingber and Folkman, 1989c; Ingber et al., 1993; Ingber, 1993; Wang et al., 1993). In these structures, a series of rigid compression-resistant struts are pull up and open through interconnection with a continuous series of tension elements. The pertinent findings obtained in studies with these "cell" models suggest that changes in ECM mechanics or cell-ECM binding interactions would produce instantaneous changes in the balance of forces that are distributed across all cytoskeletal supports (Ingber, 1991; Ingber, 1993). Furthermore, they show that a change in this force balance will result in reorientation, rather than simple deformation, of all internal support elements including the nuclear matrix (Sims et al., 1992; Ingber et al., 1993; Wang et al., 1993).

The tensegrity cell models predict that cell and nuclear shape will spread in a coordinated manner if structural and tensional continuity are maintained throughout the nuclear matrix-cytoskeletal lattice (Ingber and Jamieson, 1985). In fact, this coupling between cell and nuclear shape has been observed in both living cells (Ingber et al., 1987; Ingber and Folkman, 1989c; Ingber, 1990) and membrane-permeabilized cells (Sims et al., 1992). This latter study clearly demonstrates that mechanical tension is generated within contractile microfilaments, transmitted across integrin receptors, and resisted by external ECM binding sites. Furthermore, in membrane-permeabilized cells, one can clearly show that altering this force balance results in coordinated changes in cell, cytoskeletal, and nuclear form. Finally, recent continuum mechanics analysis of both living cells and stick and string tensegrity models has provided direct experimental evidence to support the concept that cells utilize tensegrity architecture to organize their cytoskeleton (Wang et al., 1993).

Use of tensegrity by cells is significant because it suggests that cells possess a mechanism for rapidly transmitting mechanical signals throughout the entire cytoskeleton and nucleus. Furthermore, if a mechanical signaling system exists that is based on tensegrity, then different internal regulatory signals will result depending on the mechanics of the cell's ECM. Understanding this mechanism may therefore be critical for future design scientists and tissue engineers.

INTEGRINS AS MECHANOCHEMICAL TRANSDUCERS

Taken together, these findings suggest that ECM molecules may regulate cell responsiveness to soluble growth factors and thereby, control tissue development based on their ability to act as mechanochemical transducers. Occupancy and clustering of transmembrane integrin receptors induce focal adhesion formation, recruit chemical signaling molecules, and activate gene transcription (Plopper et al., 1991; Hansen et al., 1991). Increased generation of these signals appears to be sufficient for maintenance of differentiation-specific functions. However, it is not sufficient for growth. For cells to proliferate, ECM must both bind integrins and physically resist cell-generated forces that are applied to those receptors (Ingber and Folkman, 1989b; Ingber, 1990). Exogenous mechanical forces may similarly alter cell function by changing this cellular force balance (Ingber, 1991).

Key to mechanochemical transduction is the fact that mechanical stresses are only transmitted over structural elements that are physically interconnected. Thus, ECM receptors may experience physical forces that nearby soluble molecules can not recognize. Once inside the cell, lateral transmission of physical forces through cytoskeletal interconnections in the cell cortex may alter the function of other transmembrane proteins, such as stretch-activated ion channels (Sachs, 1989) as well as other membrane-associated signaling moleucles. These forces also may regulate cytoskeletal filament assembly since polymerization of actin and tubulin have been shown to be sensitive to tension (Hill, 1981; Buxbaum and Heidemann, 1988).

As described above, the importance of the tensegrity model is that it suggests that a structural system may exist in cells that can provide direct communication of mechanical signals and harmonic information from the cell surface to the nucleus (Ingber and Jamieson, 1985; Ingber and Folkman, 1989c; Pienta and Coffey, 1991; Ingber et al., 1993). This is important because recent studies from a variety of laboratories suggest that the nuclear protein matrix is itself a structural entity which contains fixed sites for DNA replication and transcription and that it plays a critical role in chromatin packaging (Pienta et al., 1991). Interestingly, nuclear pores also physically associate with the nuclear matrix and both nuclear spreading and increased transport of proteins through nuclear pores appear to be prerequisites for entry into S phase (Yen and Pardee, 1983; Moser et al., 1983; Nicolini et al., 1986; Ingber et al., 1987).

Thus, tension-dependent changes of nuclear structure could alter nuclear functions, such as DNA synthesis and gene transcription, by changing the arrangement of associated nuclear regulatory proteins, releasing mechanical restraints to DNA unwinding, or by changing the size of channels within nuclear pores (Ingber and Jamieson, 1985; Hansen and Ingber, 1992; Ingber et al., 1993). These possibilities are supported by work from a many laboratories. For example, DNA unwinding has been shown to be rate-limiting for intiation of DNA replication (Roberts and Urso, 1988). Also, both the diameter of nuclear pores and nuclear transport rates are increased in cells that are spread and proliferating compared with round cells within confluent monolayers (Feldherr and Akin, 1990). Other possible mechanisms by which stress-induced cytoskeletal remodeling might alter cell function and gene expression have been recently described elsewhere (Ingber et al., 1993; Ingber, 1993; Wang et al., 1993).

Taken together, these observations are consistent with the concept that the effects of the ECM attachment scaffolding on cell form and function are based on a biomechanical mechanism that involves force-dependent changes in organization of the cytoskeleton and nucleus. However, these mechanical changes must act in the correct chemical context (i.e., when integrin and growth factor-dependent chemical signaling pathways are activated) in order to influence cell growth. Just as an example, while cell spreading promotes progression through G_1 phase of the cell cycle, it has little effect when exerted in G_2 (see Ingber and Folkman, 1989c for more in depth discussion).

CONCLUSION

To conclude, I would like to suggest that ECMs may be important in tissue development because they are in a critical position to transmit chemical and mechanical signals across the cell surface. Perhaps more importantly, they also provide a mechanism to integrate both chemical signaling pathways inside the cell. Thus, any effective artificial ECM must be designed with specific chemical and mechanical determinants in mind. In real terms, our work suggests that highly adhesive, rigid scaffoldings should be used when high levels of cell proliferation are required (e.g., for wound closure). Alternatively, either a highly adhesive malleable foundation or a weakly adhesive rigid substratum would be appropriate when quiescence and specialized functions are needed. Furthermore, a single rigid scaffolding that is both high adhesive and bioerodible (i.e., that will lose rigidity over time) might provide both functions: first growth would be stimulated, then upon dissolution and retraction of the broken scaffolding, the expanded population would be induced to quiesce and differentiate locally. Again, adhesivity could be varied using any one of a variety of physiological cell attachment ligands that bind functionally significant transmembrane receptors (e.g., integrins). Understanding the fine points of this type of mechanochemical control mechanism should result in definition of new rules of tissue regulation that act at the cell and molecular level. We then may be able to use these rules to define more rational design criteria and thus facilitate development of more effective and responsive artificial tissues and organs in the future.

ACKNOWLEDGEMENTS

This chapter is essentially a transcript of a lecture presented at the 1992 Keystone Symposium on Tissue Engineering. This work was supported by grants from NIH (CA-45548) and Neomorphics Inc. The liver studies were carried out in collaboration with Drs. J. Vacanti and R. Langer. Dr. Ingber is a recipient of a Faculty Research Award from American Cancer Society.

REFERENCES

Albelda SM and Buck CA (1990): Integrins and other cell adhesion molecules. *FASEB J* 4: 2868-2880.

Bell E, Ivarsson B and Merrill C (1979): Production of a tissue-like structure by contraction of collagen lattices by human fibroblasts of different proliferative potentials in vitro. *Proc Natl Acad Sci USA* 76: 1274-1278.

Ben-Ze'ev A, Robinson GS, Bucher NL and Farmer SR (1988): Cell-cell and cell-matrix interactions differentially regulate the expression of hepatic and cytoskeletal genes in primary cultures of rat hepatocytes. *Proc Natl Acad Sci USA* 85: 1-6.

Bissell DM, Arenson DM, Maher JJ and Roll FJ (1987): Support of Cultured Hepatocytes by a Laminin-rich Gel. *Am Soc Clin Inv* 79: 801-812.

Buxbaum RE and Heidemann SR (1988): A thermodynamic model for force integration and microtubule assembly during axonal elongation. *J Theor Biol* 134: 379-390.

Cima LG, Vacanti JP, Vacanti C, Ingber DE, Mooney DJ and Langer RL (1991): Tissue Engineering by Cell Transplantation Using Biodegradable Polymers. *J Biomech Eng* 38: 145-158.

Dunn JC, Yarmush ML, Koebe HG and Tompkins RG (1989): Hepatocyte Function and Extracellular Matrix Geometry: Long-term culture in a Sandwich Configuration. *FASEB J* 3: 174-177.

Emerman JT and Pitelka DR (1977): Maintenance and induction of morphological differentiation in dissociated mammary epithelium on floating collagen membranes. *In Vitro Cell Dev Biol* 13: 316-328.

Erickson CA and Trinkaus JP (1976): Microvilli and blebs as sources of reserve surface membrane during cell spreading. *Exp Cell Res* 99: 375-384.

Feldherr C and Akin D (1990): The permeability of the nuclear envelope in dividing and nondividing cell cultures. *J Cell Biol* 98: 1973-1984.

Hansen LH, Mooney DJ and Ingber DE (1991): Clustering of integrins induces expression of the early growth response gene, junB, in hepatocytes, independent of cell spreading. *J Cell Biol* 115: 443a.

Hansen LK and Ingber DE (1992): Regulation of nucleocytoplasmic transport by mechanical forces transmitted through the cytoskeleton. In: Feldherr C, ed. *Nuclear Trafficking*, Academic Press: Orlando, FL pp. 71-86.

Harris AK (1982): Traction, and its relations to contraction in tissue cell locomotion. In: *Cell Behavior*, (Bellairs, R., Curtis, A., Dunn, G., eds.) Cambridge University Press: Cambridge, pp. 109-134.

Hill TL (1981): Microfilament or microtubule assembly or disassembly against a force. *Proc Natl Acad Sci USA* 78: 5613-5617.

Ingber DE (1990): Fibronectin controls capillary endothelial cell growth by modulating cell shape. *Proc Natl Acad Sci USA* 87: 3579-3583.

Ingber DE (1991). Integrins as mechanochemical transducers. *Curr Opin Cell Biol* 3: 841-848.

Ingber DE (1993) Cellular tensegrity: defining new rules of biological design that govern the cytoskeleton. *J Cell Sci* - in press.

Ingber DE and Folkman J (1989a): Mechanochemical switching between growth and differentiation during fibroblast growth factor-stimulated angiogenesis *in vitro*: role of extracellular matrix. *J Cell Biol* 109: 317-330.

Ingber DE and Folkman J (1989b): How does extracellular matrix control capillary morphogenesis? *Cell* 58: 803-805.

Ingber DE and Folkman J (1989c) Tension and compression as basic determinants of cell form and function: utilization of a cellular tensegrity mechanism. In: Stein W, Bronner F, eds. *Cell Shape: Determinants, Regulation and Regulatory Role.* Academic Press: Orlando, pp. 1-32.

Ingber DE and Jamieson JD (1985): Cells as tensegrity structures: architectural regulation of histodifferentiation by physical forces tranduced over basement membrane. In: Andersson LC, Gahmberg CG, Ekblom P, eds. IN: *Gene Expression During Normal and Malignant Differentiation*. Academic Press: Orlando, pp. 13-32.

Ingber DE, Karp S, Plopper G, Hansen L and Mooney D (1993): Mechanochemical transduction across extracellular matrix and through the cytoskeleton. In: Frangos JA, Ives CL, eds. *Physical Forces and the Mammalian Cell*. Academic Press: San Diego. pp. 61-78.

Ingber DE, Madri JA and Folkman J (1987): Extracellular matrix regulates endothelial growth factor action through modulation of cell and nuclear expansion. *In Vitro Cell Dev Biol* 23: 387-394.

Ingber DE, Prusty, D, Frangione J, Cragoe EJ Jr, Lechene C and Schwartz M (1990): Control of intracellular pH and growth by fibronectin in capillary endothelial cells. *J Cell Biol* 110: 1803-1812.

Koch J (1917): The laws of bone architecture. *Am J Anat* 21: 177-198.

Kreis TE and Birchmeier W (1980): Stress fiber sarcomeres of fibroblasts are contractile. *Cell* 22: 555-561.

Li ML, Aggeler J, Farson DA, Hatier C, Hassell J and Bissel MJ (1987): Influence of a reconstituted basement membrane and its components on casein gene expression and secretion in mouse mammary epithelial cells. *Proc Natl Acad Sci USA* 84: 136-140.

McNamee H, Ingber D and Schwartz M (submitted): Intersection of growth factor and extracellular matrix-activated signaling pathways.

Michalopoulos G and Pitot HC (1975): Primary Culture of Parenchymal Liver Cells on Collagen Membranes. *Exp Cell Res* 94: 70-78.

Mochitate K, Pawelek P and Grinnell F (1991): Stress relaxation of contracted collagen gels: disruption of actin filament bundles, release of cell surface fibronectin, and down-regulation of DNA and protein synthesis. *Exp Cell Res* 193: 198-207.

Mooney D, Hansen L, Farmer S, Vacanti J, Langer R and Ingber D (1992): Switching from differentiation to growth in hepatocytes: control by extracellular matrix. *J Cell Physiol* 151: 497-505.

Mooney DJ, Langer R, Hansen LK, Vacanti JP and Ingber DE (1992): Induction of hepatocyte differentiation by the extracellular matrix and an RGD-containing synthetic peptide. *Proc Mat Res Soc Symp Proc* 252: 199-204.

Moser GC, Fallon RJ and Meiss HK (1981): Fluorometric measurements and chromatin condensation patterns of nuclei from 3T3 cells throughout G1. *J Cell Physiol* 106: 293-301.

Nicolini C, Belmont AS and Martelli A (1986): Critical nuclear DNA size and distribution associated with S phase initiation. *Cell Biophys* 8: 103-117.
Opas M (1989): Expression of the differentiated phenotype by epithelial cells *in vitro* is regulated by both biochemistry and mechanics of the substratum. *Dev Biol* 131: 281-293.
Pienta KJ, Getzenberg RH and Coffey DS (1991): Cell Structure and DNA Organization. *Crit Rev Eukary Gene Express* 1: 355-385.
Pienta KJ and Coffey DS (1991): Cellular harmonic information transfer through a tissue tensegrity-matrix system. *Med Hypoth* 34: 88-95.
Plopper G, Schwartz MA, Chen LB, Lechene C and Ingber DE (1991): Binding of fibronectin induces assembly of a chemical signaling complex on the cell surface. *J Cell Biol* 115: 130a.
Reid LM (1990): Stem Cell Biology, Hormone/Matrix Synergies and Liver Differentiation. *Curr Opin Cell Biol* 2: 121-130.
Roberts JM and D'Urso G (1988): An origin unwinding activity regulates initiation of DNA replication during mammalian cell cycle. *Science* 241: 1486-1489.
Sachs F (1989): Ion channels as mechanical transducers. In: Stein W, Bronner F, eds. *Cell Shape: Determinants, Regulation and Regulatory Role.* Academic Press: Orlando, pp. 63-92.
Sawada N, Tomomura A, Sattler CA, Sattler GL, Kleinman HK and Pitot HC (1987): Effects of Extracellular Matrix Components on the Growth and Differentiation of Cultured Hepatoctyes. *In Vitro Cell Dev Biol* 23: 267-273.
Schwartz MA (1992): Transmembrane signalling by integrins. *Trends Cell Biol* 2: 304-308.
Schwartz MA, Ingber DE, Lawrence M, Springer TA and Lechene C (1991a): Multiple Integrins Share the Ability to Induce Elevation of Intracellular pH. *Exp Cell Res* 195: 533-535.
Schwartz MA, Lechene C and Ingber DE (1991b): Insoluble Fibronectin Activates the $Na+/H+$ Antiporter by Clustering and Immobilizing Integrin $\alpha 5\beta 1$, independent of cell shape. *Proc Natl Acad Sci USA* 88: 7849-7853.
Sims J, Karp S and Ingber DE (1992): Altering the cellular mechanical force balance results in integrated changes in cell, cytoskeletal, and nuclear shape. *J Cell Sci* - in press.
Stoker AW, Streuli CH, Martins-Green M and Bissell MJ (1990): Designer Microenvironments for the Analysis of Cell and Tissue Function. *Curr Opin Cell Biol* 2: 864-874.
Vacanti JP, Morse M, Saltzman W, Domb A, Perez-Atayde A and Langer R (1988): Selective cell transplantation using bioabsorbable artificial polymers as matrices. *J Pediatr Surg* 23: 3-9.

Wang N, Butler JP and Ingber DE (submitted): Probing the molecular basis of mechanotransduction across the cell surface and through the cytoskeleton.
Yen A and Pardee AB (1979): Role of nuclear size in cell growth initiation. *Science* 204: 1315-1317.

MODULATION OF CARDIAC GROWTH BY SYMPATHETHIC INNERVATION: DIFFERENTIAL RESPONSE BETWEEN NORMOTENSIVE AND HYPERTENSIVE RATS

Dianne L. Atkins
Department of Pediatrics
University of Iowa College of Medicine

INTRODUCTION

Sympathethic innervation of the heart develops during late gestation and is not fully mature until the end of the neonatal period. The onset of neuroeffector transmission coincides with the cessation of myocyte mitosis (Claycomb, 1976) and the beginning of cellular hypertrophy (Clubb and Bishop, 1984). This process must occur during normal development and growth but also occurs pathologically in the setting of heart disease and hypertension. Although hemodynamic load is a major stimulus for cardiac hypertrophy, it is not the sole regulatory mechanism as evidenced by hypertrophic cardiomyopathy, which occurs in the presence of presumed normal hemodynamics.

The role of the sympathetic nervous system in modulation of cardiac growth has been suggested by two separate experimental models. Tucker and colleagues (Tucker and Gist, 1986; Tucker, 1990) have reported decreased external dimension and weight of fetal rat atria and ventricles implanted into the anterior eye chamber previously denervated by stellate ganglionectomy. Additionally, chronic catecholamine exposure has been shown to increase protein content of myocardial cells cultured in a serum-free media (Simpson et al., 1982). Within this system, the response is specific to $alpha_1$-adrenoceptor activation (Simpson, 1985)

We have previously described a co-culture model of neonatal rat myocardial cells harvested prior to *in vivo* innervation and sympathetic explants (Marvin et al., 1984; Atkins and Marvin, 1989). The neurons form neuromuscular junction and neuroeffector transmission develops within 48 hours of culture. This system removes neurohumoral and hemodynamic factors, providing the environment to study differences in neurotrophic effect in isolation. These studies were undertaken to understand the mechanisms of myocardial growth modulated by innervation and to compare the responses of two genetically bred rat strains, Wistar-Kyoto and the spontaneously hypertensive rat.

METHODS

The culture system has been described elsewhere (Atkins and Marvin, 1989; Atkins et al., 1991). Pertinent details included harvesting myocardial cells from rats less than 48 hours old, a mild trypsin and collagenase digestion, plating concentration of 10^5 cells/ml, and providing a media enriched with horse serum. The neuronal explants were obtained from the thoracolumbar sympathetic chains of the same rats. Cultures from WKY and SHR rats were prepared simultaneously and all experiments performed on day 4 of culture.

Cell size was measured using stereologic techniques modified for single cell analysis (Atkins et al., 1991; Atkins et al., 1993). Single myocytes were serially sectioned en face at a thickness of 120 nm and every fifth section prepared for transmission electron microscopy. Point counting was preformed with grids of 10 and 20 mm and the cumulative points used to calculate total myocyte volume, organelle volumes and relative volumes. Data were analyzed using ANOVA with randomized block design, blocking for culture day. Post hoc comparisons were made with the Bonferroni's Test for Multiple Comparisons, while controlling the overall probability of Type I error.

RESULTS

Effects of Sympathetic Innervation

Innervation did not affect the height, shape of the myocytes, nor the organization of the cytoplasmic components. However, mean cell volume of innervated WKY myocytes increased by 44% (Fig 1) with proportional increases of the nucleus and cytoplasm. The absolute volume of the mitochondria and other cellular components (defined as all structures, excluding the nucleus, mitochondria and sarcomeres) also increased (Fig 2). Interestingly, sarcomere volume of innervated cells did not differ from non-innervated cells. Relative volumes of the nucleus and mitochondria remained the same as control cells, but the sarcomeres and other cellular components were different as a result of the absolute sarcomere volume (Fig 3).

Effect of Adrenoceptor Blockade

The presumed mechanism of this neurotrophic-induced growth was adrenoceptor stimulation, as suggested by both *in vivo* and *in vitro* experiments (Laks et al., 1973; Simpson et al., 1982; Lee et al., 1988). To test this hypothesis, norepinephrine concentrations in the culture media were measured and the size of innervated cells was measured after incubation with 2mM propranolol and prazosin. These concentrations

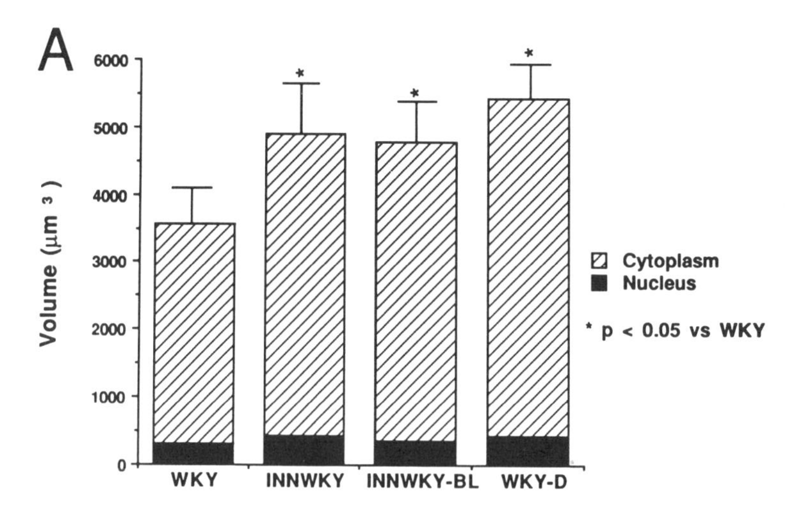

FIGURE 1. Absolute volume of WKY myocytes in culture. Error bars are SE. Abbreviations: WKY isolated myocytes; INNWKY innervated myocytes; INNWKY-BL innervated myocytes, incubated with 2 mM propranolol and 2 mM prazosin; WKY-D non-innervated myocytes in co-culture, distant from the neuronal explants. Asterisk (*) indicates P<0.05. (Reprinted with permission of Academic Press, Inc., London)

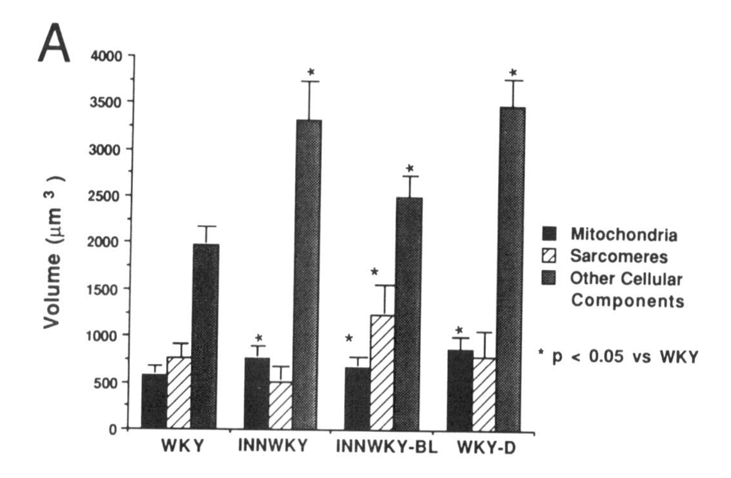

FIGURE 2. Absolute volumes of intracellular components of WKY myocytes. Abbreviations are the same as Fig 1. (Reprinted with permission of Academic Press, Inc., London)

were sufficient to prevent a change in myocyte contractile frequency when acutely exposed to norepinephrine. Norepinephrine concentrations were 10^{-11} to 10^{-12} M, 10,1000 to 100,000 fold lower than that previously reported to promote growth (Simpson et al., 1982). Additionally, adrenoceptor blockade did not attenuate the effects of innervation on myocyte, cytoplasmic or nuclear volume (Fig 1-3), indicating that these effects were independent of receptor stimulation. There were no significant differences in any of the relative volumes between the innervated myocytes and those exposed to adrenoceptor blockade (Fig 3).

Presence of Soluble Factor Within the Culture Medium

The volume of myocytes within the co-cultures, but were not innervated was 52% larger than control myocytes and was not different from innervated myocytes (Fig 1-3). Intracellular organelles increased similarly. This indicated that a soluble factor was in the culture medium of the innervated cells, but not in medium of the control cells. Because these are primary cultures, the cell population is heterogenous, with non-myocardial cells in all cultures and neurons and non-neuronal cells within the co-cultures.

Differential response of SHR myocytes

The spontaneously hypertensive rat (SHR) is a genetic counterpart to the WKY, characterized by the development of severe hypertension by three months of age. Death occurs prematurely from cardiac failure or stroke. Cardiac hypertrophy is a well-documented feature, but its development may actually precede the development of hypertension (Gray and Brown, 1989). It is possible to dissociate cardiomegaly from hypertension, suggesting that cardiomegaly is an intrinsic abnormality of the strain. Moreover, the growth response of 8 week old SHR cultured myocytes to norepinephrine is less than that of WKY myocytes (Ikeda et al., 1990). Thus, we tested the effect of innervation on SHR myocytes compared to the WKY.

Control SHR myocytes were significantly larger that WKY myocytes (Fig 4), and all intracellular organelle volumes were proportionately larger. Relative volumes did not differ between the two strains. Most interestingly, total myocyte volume, and all intracellular volumes were unchanged by sympathetic innervation. The SHR myocytes did not differ from innervated WKY myocytes.

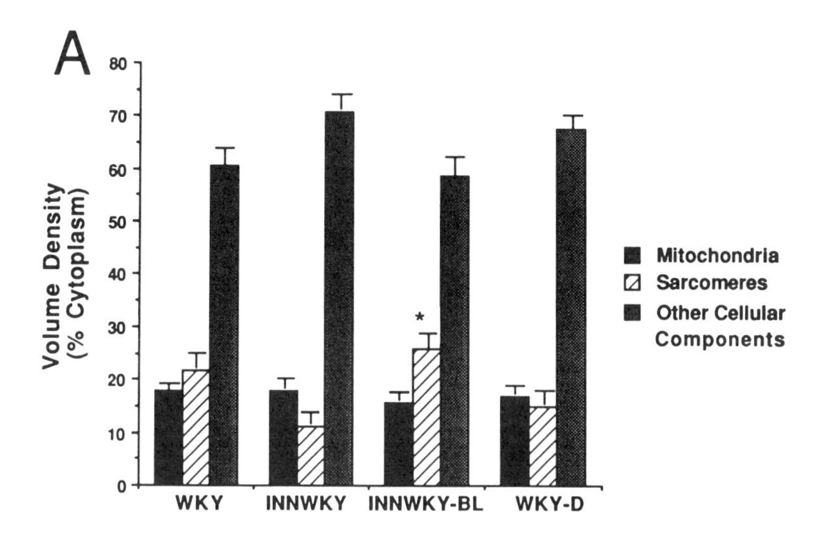

FIGURE 3. Relative volumes of intracellular components of WKY myocytes. Abbreviations are the same as Fig 1. (Reprinted with permission of Academic Press, Inc., London)

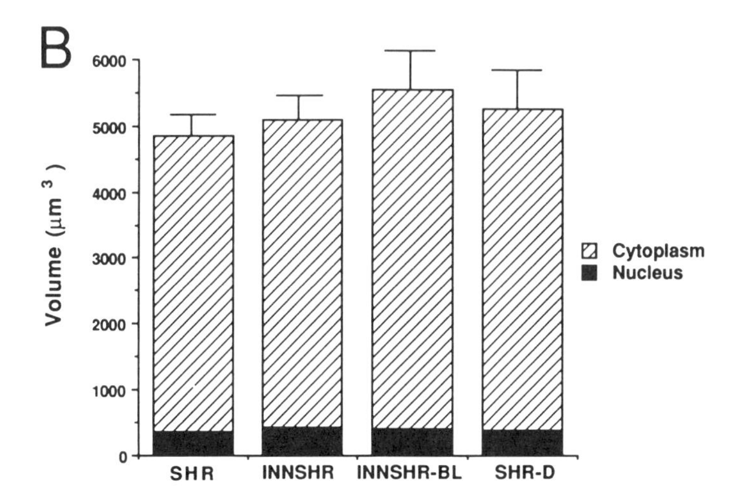

FIGURE 4. Absolute volume of SHR myocytes in culture. Abbreviations are the same as Fig 1, except myocyte and neuronal source is SHR animals. (Reprinted with permission of Academic Press, Inc., London)

DISCUSSION

Myocardial cell growth has been studied using a variety of methods, each with intrinsic suitability and shortcomings. Measurement of surface area, diameter of resuspended cells, as well as more classic stereologic techniques (Simpson et al., 1982; Delcarpio et al., 1989) has led to a marked increase in our understanding of the mechanisms which control myocardial growth. The predominant advantage of the method we developed is the precise measurement of cell size as well as the availability of additional information about ultrastructural alterations induced by experimental manipulation. Additionally, this method of identifying individual innervated and noninnervated cells in culture and relocating them for transmission electron microscopy, allowed us to perform precise volume calculation on a well-defined population of cells.

Growth of cultured WKY myocytes can be increased by sympathetic innervation which is independent of adrenoceptor stimulation or anatomic contact between the neuron and myocyte. There appears to be some factor(s) produced by the neurons, the myocytes in response to innervation or non-neuronal cells which promotes cell growth. There are reports of two proteins which promote growth of myocytes. One, myotrophin (Sen et al., 1990), has been purified from adult SHR hypertrophied hearts. The other is produced by cardiac nonmyocytes but as yet is undefined (Long et al., 1991). As there was no difference in non-myocyte number among the culture groups, this does not appear to be explain our results. A neuronal source is a strong contender for the soluble factor. This is supported by the presence of a soluble factor in conditioned media from a neuronal cell line (PC12 cells) which promotes increased myocyte contractile function (Green et al., 1990) and increased myocyte uptake of $[^{35}]$S-methionine (unpublished observations).

The larger SHR cells prior to innervation and lack of response to innervation was an unexpected finding. These results imply strain differences not related to hemodynamic load. The differences may exist in the maturation rate of the neurons (McCarty et al., 1987) or the myocytes (Engelmann and Gerrity, 1988; Engelmann et al., 1989). Another alternative is the existence of a "window" of responsiveness when myocytes respond to regulatory factors which control growth, the timing of which may be different between the WKY and SHR.

The interaction of neonatal myocytes and autonomic innervation is a field of active investigation. Several laboratories have now demonstrated that neurons have a major effect on the myocytes as a result of changes in the myocyte, i.e. G protein acquisition (Steinberg et al., 1985), changes in sodium and calcium channels (Zhang et al., 1992; Ogawa et al., 1992) and changes in contraction frequency which are independent of spontaneous firing of the neurons, i.e. post-synaptic

(Drugge et al., 1985, Atkins and Marvin, 1989). Together these papers indicate a major developmental role of autonomic innervation in the regulation of immature myocytes. This model system is particularly suited to studying this process because of the capacity to accelerate sympathetic innervation compared to *in vivo* innervation, which is not complete until several days to weeks after birth. Further inquiry will permit increased understanding of neurohumoral, genetic programming and/or developmental regulation of cell function. Strain or species differences are especially amenable to this type of study and, as illustrated by the differences, between WKY and SHR clearly need to be examined. In addition to the pharmacologic and biochemical studies, this model will assist in the investigation of gene expression and gene transcription.

ACKNOWLEDGEMENTS

This research was supported by grants HL35600 and Hl43273 from the National Heart, Lung and Blood Institute and by the Iowa Affiliate, American Heart Association.

REFERENCES

Atkins DL, Rosenthal JK, Krumm PA and Marvin WJ,Jr. (1991): Application of stereological analysis of cell volume to isolated myocytes in culture with and without adrenergic innervation. *Anat Rec* 231: 209-217.

Atkins DL, Rosenthal JK, Krumm PA and Marvin WJ,Jr. (1993): Differential growth of neonatal WKY and SHR ventricular myocytes within sympathetic co-cultures. *J Mol Cell Cardiol* In press.

Atkins DL and Marvin WJ,Jr. (1989): Chronotropic responsiveness of developing sinoatrial and ventricular rat myocytes to autonomic agonists following adrenergic and cholinergic innervation in vitro. *Circ Res* 64: 1051-1062.

Claycomb WC (1976): Biochemical aspects of cardiac muscle differentiation. Possible control of deoxyribonucleic acid synthesis and cell differentiation by adrenergic innervation and cyclic adenosine 3':5'-monophosphate. *J Biol Chem* 251: 6082-6089.

Clubb FJ and Bishop SP (1984): Formation of binucleated myocardial cells in the neonatal rat. An index for growth and hypertrophy. *Lab Invest* 50: 571-577.

Delcarpio JB, Claycomb WC and Moses RL (1989): Ultrastructural morphometric analysis of cultured neonatal and adult rat ventricular cardiac muscle cells. *Am J Anat* 345: 335-345.

Drugge ED, Rosen MR and Robinson RB (1985): Neuronal regulation of the development of the α-adrenergic chronotropic response in the rat heart. *Circ Res* 57: 415-423.

Engelmann GL, Boehm KD, Haskell JF, Khairallah PA and Ilan J (1989): Insulin-like growth factors and neonatal cardiomyocyte development: Ventricular gene expression and membrane receptor variations in normotensive and hypertensive rats. *Mol Cell Endocrinol* 63: 1-14.

Engelmann GL and Gerrity RG (1988): Biochemical characterization of neonatal cardiomyocyte development in normotensive and hypertensive rats. *J Mol Cell Cardiol* 20: 169-177.

Gray SD and Brown H (1989): Reciprocal embryo transfer between SHR and WKY: II. Effect on cardiovascular development. *Clin Exp Hypertens* 13: 963-969.

Green SH, Glover RM, Krumm PA and Atkins DL (1990): Neuronal regulation of cardiac myocyte contractility concomitant with regulation of expression of a 23 kDa protein. *Circ* 82 (Suppl): 746.

Ikeda U, Tsuruya Y, Tsuyoshi K, Oguchi A, Oohara T and Yaginuma T (1990): Ventricular cells in culture from adult spontaneously hypertensive rat exhibit decreased growth. *J Hypertens* 8: 1003-1006.

Laks MM, Morady F and Swan HJC (1973): Myocardial hypertrophy produced by chronic infusion of subhypertensive doses of norepinephrine in the dog. *Chest* 64: 75-78.

Lee HR, Henderson SA, Reynolds R, Dunnmon P, Yuan D and Chien KR (1988): α_1-adrenergic stimulation of cardiac gene transcription in neonatal rat myocardial cells. *J Biol Chem* 263: 7352-7358.

Long CS, Henrich CJ and Simpson PC (1991): A growth factor for cardiac myocytes is produced by cardiac nonmyocytes. *Cell Regul* 2: 1081-1095.

Marvin WJ,Jr., Atkins DL, Chittick VL, Lund DD and Hermsmeyer K (1984): *In vitro* adrenergic and cholinergic innervation of the developing rat myocyte. *Circ Res* 55: 49-58.

McCarty R, Kirby RF, Cierpial MA and Jenal TJ (1987): Accelerated development of cardiac sympathetic responses in spontaneously hypertensive (SHR) rats. *Behav Neural Biol* 48: 321-333.

Ogawa S, Barnett JV, Sen L, Galper JB, Smith TW and Marsh JD (1992): Direct contact between sympathetic neurons and rat cardiac myocytes in vitro increases expression of functional calcium channels. *J Clin Invest* 89: 1085-1093.

Sen S, Kundu G, Mekhail N, Castel J, Misono K and Healy B (1990): Myotrophin: Purification of a novel peptide from spontaneously

hypertensive rat heart that influences myocardial growth. *J Biol Chem* 265: 16635-16643.
Simpson P, McGrath A and Savion S (1982): Myocyte hypertrophy in neonatal rat heart cultures and its regulation by serum and by catecholamines. *Circ Res* 51: 787-801.
Simpson P (1985): Stimulation of hypertrophy of cultured neonatal rat heart cells through the α_1-adrenergic receptor and induction of beating through an α_1- and β_1-adrenergic receptor interaction. Evidence for independent regulation of growth and beating. *Circ Res* 56: 884-894.
Steinberg SF, Drugge ED, Bilezikian JP and Robinson RB (1985): Acquisition by innervated cardiac myocytes of a pertussis toxin-specific regulatory protein linked to the α1-receptor. *Science* 230: 186-188.
Tucker DC (1990): Genetic, neurohumoral, and hemodynamic influences on spontaneously hypertensive rat heart development in oculo. *Hypertension* 15: 247-256.
Tucker DC and Gist R (1986): Sympathetic innervation alters growth and intrinsic heart rate of fetal rat atria maturing *in oculo. Circ Res* 59: 534-544.
Zhang J-F, Robinson RB and Siegelbaum SA (1992): Sympathetic neurons mediate developmental change in cardiac sodium channel gating through long-term neurotransmitter action. *Neuron* 9: 97-103.

LIVER SUPPORT THROUGH HEPATIC TISSUE ENGINEERING

Mehmet Toner[†], Ronald G. Tompkins[†], and Martin L. Yarmush[†,††]

[†] Surgical Services, Massachusetts General Hospital, and
Shriners Burns Institute
Boston, MA 02114, and
[††] Department of Chemical and Biochemical Engineering
Rutgers University
Piscataway, NJ 08854.

INTRODUCTION

Tissue engineering is a relatively new field which brings together several disciplines, including bioengineering, molecular and cellular biology and clinical medicine, with the ultimate goal of producing tissue or organ substitutes. The spectrum of applications of tissue engineering is indeed wide from artificial skin substitutes (Bell et al., 1991) and transplantable cell-polymer matrices (Cima et al., 1991) to bioartificial organs such as the bioartificial liver and bioartificial pancreas (Yarmush et al.,1992a; 1992b; Colton and Avgoustiniatos, 1992). All of these technologies have a number of common aspects, such as (1) the identification of reliable and efficient sources of cells and isolation of cells from tissues, (2) means of culture and preservation of cells, and (3) tissue design.

In this presentation, we focus on our recent hepatic tissue engineering studies, the aim of which is to develop an extracorporeal liver support device. Our presentation is divided into three sections: (1) studies on the isolation procedures for rat and human hepatocytes, (2) methods for long-term culture and preservation of hepatocytes, and (3) a discussion of important issues in the design of bioartificial liver support devices. This paper is not meant to be an exhaustive review of the literature but rather an outline of our laboratory's recent efforts in this area.

SOURCE AND ISOLATION OF HEPATOCYTES

One of the major challanges in cell-based therapies is the efficient procurement of cells or tissue which is needed. In the case of bioartificial liver, there are a number of different sources for hepatocytes. These sources include animal cells, human cells, and hepatoma cell lines. Although some hepatoma cell lines have been shown to exhibit many crucial hepatospecific functions (Knowles et al., 1980), the clinical use of these cells may be limited given their tumor origin. Cells from animal and human sources are another alternative which requires the development of efficient procedures to isolate hepatocytes from intact livers or liver slices.

Isolation Procedures for Rat and Human Livers

Since the introduction by Berry and Friend (1969), the collagenase liver perfusion has been widely accepted as the standard method for the isolation of viable hepatocytes. Seglen (1976) extensively studied and optimized this process for hepatocyte isolation and his protocol has since been modified by others (Kraemer et al., 1986). Figure 1 schematically depicts the standard procedure used for rat hepatocyte isolation in our laboratory based on the modified Seglen procedure. The liver is first perfused with calcium free Krebs Ringer bicarbonate buffer at 35 ml/min. The perfusate is maintained at 37°C and equilibrated with 95% O_2 and 5% CO_2. The liver is subsequently perfused with a 0.05% collagenase solution for 10 min at the same flow rate. The resulting cell suspension is then purified to obtain hepatocytes using a Percoll gradient technique. Routinely, 200-300 million cells are isolated from an 8-g liver, with viabilities ranging from 90% to 95%. Nonparenchymal cells, as judged by their size (< 10 μm in diameter) and morphology, are usually less than 1% of the purified population. With an 8 gram liver for a typical adult rat, and 100×10^6 parenchymal cells per gram of liver tissue, one can expect about 800×10^6 parenchymal cells to be present in the rat liver. Therefore, the protocol used for the isolation of rat hepatocytes results in a yield of, at best, 25-40% of the total number of hepatocytes available. Although this level of inefficiency is not problematic on a laboratory scale, it can become a limiting step for clinical application when large number of cells are needed.

The isolation protocol for human hepatocytes represents an even a greater challenge. Relatively little is known about the separation of human liver into single cell suspensions. The initial approach taken to isolate single cell suspensions from whole livers and liver pieces was described by Reese and Byard (1981) in which they perfused human liver slices in a similar fashion to that used to perfuse whole livers in rats. Following this approach, we developed a pressure-controlled system to digest human liver tissue into single cell suspensions by cannulation and perfusion of liver pieces taken from surgical hepatic resections (Ryan et al., 1993). In order to equally perfuse each cannulated vessel in a given liver piece, the perfusions are pressure-controlled as opposed to rate-controlled. Although the average yield of viable liver cells per gram of liver was increased from 1.1×10^7 for rate-controlled procedure to 4.7×10^7 for a controlled perfusion pressure of 112 mmHg, the procedure still yields only a fraction of cells which are available in the liver pieces. Thus, the need to improve the isolation procedure is crucial for the clinical applications of the bioartificial liver device using human hepatocytes.

Molecular and Macroscopic Level Analyses of Liver Perfusion

In addition to high yield isolations, it is desirable to have hepatocytes which function maximally immediately after the isolation procedure. In reality, cell function, including hepatic protein secretion, lag up to a week after isolation before reaching a steady-state maximal rate (Dunn et al., 1991a). If this lag is due to the initial isolation procedure, it may be possible to reduce or eliminate this lag by altering the isolation process. In order to alter the process, however, we must first understand which physicochemical parameters(s) result in damage.

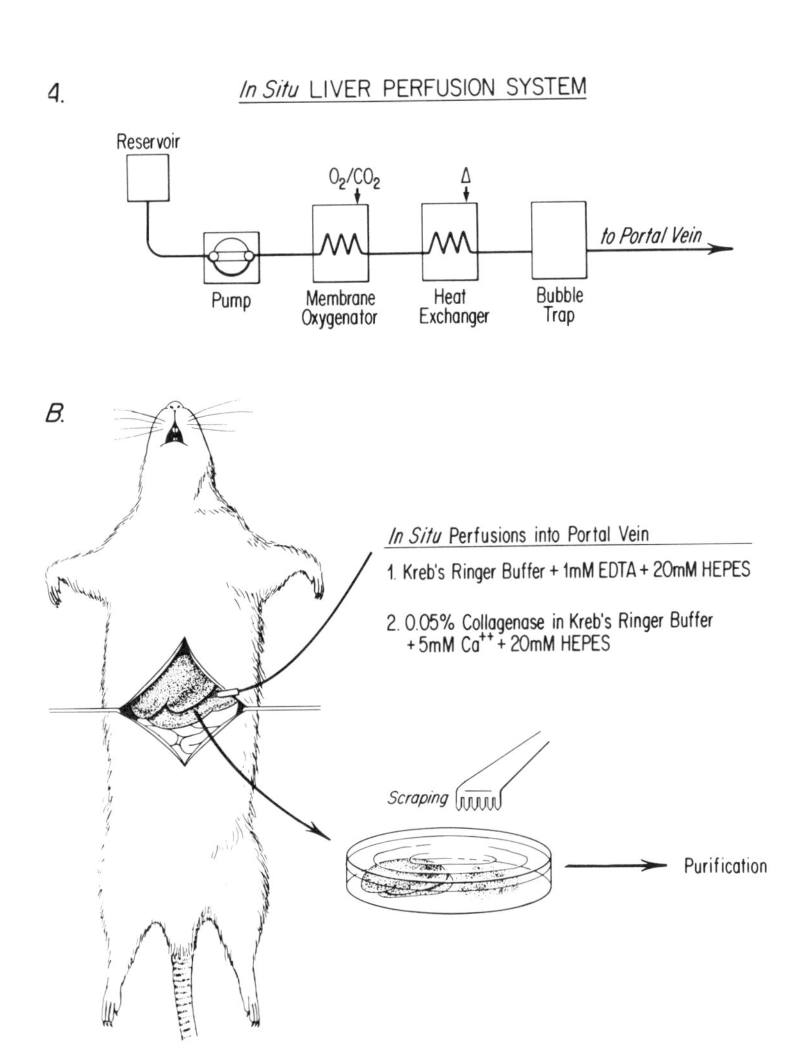

FIGURE 1. Schematic diagram of the procedure used for isolating rat hepatocytes.

Using polyribosome size and albumin mRNA as indicators, experiments were performed to evaluate several physicochemical parameters of the isolation protocol (Dunn et al., 1992b). Immediately after isolation, it was found by density centrifugation that the larger polyribosomes which are present in normal liver tissue were distrupted into smaller polyribosomes. This result was further confirmed by the analysis of albumin mRNA content which showed the distribution of albumin mRNA associated polyribosomes had also shifted from larger to smaller polyribosomes. The calculated average size of an albumin polyribosome was 5 for the freshly isolated cells, compared to 9 for the normal liver. Furthermore, the hepatocytes isolated by the standard technique did not show a restoration of the larger polyribosome pattern in culture for many days. In an initial attempt to understand which process parameters were causing this change in the cells, we analyzed the cell isolation process, paying particular attention to perfusate composition, flow rate, and temperature. By changing these parameters, we demonstrated that the reduction in large polyribosome size can be limited or avoided by the use of more complex perfusate, a more physiological flow rate, or a lower perfusion temperature. The two strategies that resulted in the near or total preservation of polyribosome size were: (1) perfusion with Williams' media E at 10 ml/min, and (2) perfusion with the usual buffer at room temperature. These modifications, however, did not lead to high yield isolations of viable hepatocytes. It is clear from these studies that further investigation is needed before we can design gentle yet efficient isolation protocols.

Most recently, we have studied perfused livers using NMR techniques. Figure 2 shows the change in the levels of ATP during the standard perfusion procedure at a flow rate of 35 ml/min. Minimal changes were observed during the control perfusion before and after EDTA was added. However, a major drop in ATP concentration and pH occurred shortly after collagenase was added to the perfusate. To evaluate the effect of collagenase on the liver, we also monitored the perfusion in real time using NMR imaging. Our results showed that organ perfusion was compromised immediately after the collaganese perfusion. It is plausible that the ATP and pH changes were caused by oxygen limitations when collagenase distrupts convective flow within the sinusoids. Alternative approaches which would allow more uniform tissue perfusion and, thus, exposure of the tissue to the collagenase are needed. One such approach would involve the perfusion of the tissue with collagenase at 4°C to initially reduce or eliminate the enzymatic activity. Then the temperature could be rapidly increased to 37°C to activate the enzyme which would be uniformly distributed throughout the tissue. Other possibilities may also exist and further studies to optimize the large-scale isolation of cells from organs and tissues are needed.

LONG-TERM CULTURE OF HEPATOCYTES AND PRESERVATION OF HEPATOCELLULAR FUNCTION

Effect of the Extracellular Matrix Geometry on Hepatocyte Function and Morphology

Hepatocytes in vivo have a belt of apical surface dividing two basolateral surfaces that are in contact with the extracellular matrix. In an effort

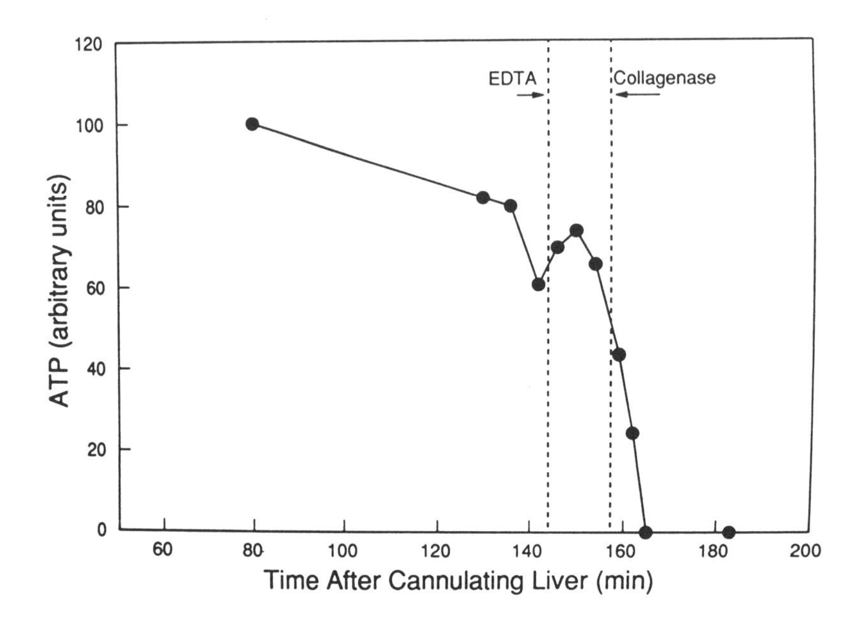

FIGURE 2. Time course of ATP changes in perfused liver during isolation process as measured by NMR.

to mimic this in vivo liver geometry, we examined the effect of sandwiching a monolayer of hepatocytes between two layers of hydrated rat tail collagen matrix (Dunn et al., 1989;1991;1992a). A wide range of liver-specific functions was evaluated comparing the sandwich and the single layer culture techniques. In general, hepatocytes cultured in the sandwich system maintained hepatospecific function quantitatively; whereas, hepatocytes on a single layer of collagen ceased functioning in one to two weeks. Table 1 summarizes the cellular parameters evaluated and cites the appropriate references.

The rates of protein secretion measured in the sandwich system were comparable to those reported for the normal liver. The rate of albumin, transferin, and fibrinogen have been estimated at 0.74, 0.14, and 0.06 mg h^{-1} (g of liver)$^{-1}$, respectively (Jeejeebhoy et al., 1973; Morgan and Peters, 1971; Peters, 1985). If one assumes $100x10^6$ hepatocytes in 1-g of liver, these synthesis rates can be converted to 7.4, 1.4, and 0.6 μg h^{-1} 10^{-6} cells, which are rates that are comparable to the rates of protein secretion observed in the sandwich culture system.

Although, the single layer culture system showed only transient protein secretion and secretion rates orders-of-magnitude less than the in vivo rates, this drop in function could be reversed during the first seven days of culture (Dunn et al., 1989;1991). The addition of the second layer of collagen matrix to the single layer cultures reconfigured the hepatocytes both functionally and histologically.

TABLE I. Characteristics of hepatocytes examined in the sandwich culture technique.

Characteristic Examined	Reference
Protein Secretion	Dunn et al., 1989; 1991; 1992a
Bile Salt Secretion	Dunn et al., 1991
Urea Secretion	Dunn et al., 1991
Cytochrome P-450 Activity	unpublished data
Oxygen Uptake	Rotem et al., 1992
Proline Effect	Lee et al., 1992; 1993
Cryopreservation	Koebe et al., 1990;
	Borel Rinkes et al., 1992a; 1992b
	Toner et al., 1992
	Hubel et al., 1991
	Harris et al., 1991
Cytoskeletal Dynamics	Ezzell et al., 1993
Cell Shape	Ezzell et al., 1993
	Rotem et al., 1992
Membrane Protein Position	Dunn, 1992
	Yarmush et al., 1992b

The exact mechanism by which the double gel transmits its effect is not known, however, it appears that at least part of the regulation occurs at the transcriptional level (Dunn et al., 1992a). Our studies have shown that sandwiched hepatocytes exhibit significantly higher transcriptional activity than cells on a single collagen layer. No significant differences between single layer and sandwich cultures were observed with respect to their translational activities. There is some evidence that the collagen gel overlay may act as a structural barrier which retains newly synthesized extracellular matrix materials. Further evidence in support of this hypothesis was obtained from our studies on proline supplementation and will be discussed in the next section.

In addition to functional assessment of sandwiched hepatocytes, experiments were performed to characterize the importance of the extracellular matrix configuration in regulating cytoskeletal protein organization and expression (Ezzell et al., 1993). Sandwiched hepatocytes are polygonal in shape and morphologically resemble cells from intact liver. Actin filaments (F-actin) in these hepatocytes are concentrated under the plasma membrane in regions of cell-cell contact. In contrast, hepatocytes cultured on a single collagen layer are flattened and motile and have F-actin containing stress fibers. Microtubules formed an interwoven network in hepatocytes cultured in a sandwich gel, but in single gel cultures they formed long parallel arrays extending out to the cell periphery. Both actin and tubulin mRNA levels were several fold greater in hepatocytes cultured on a single layer. Time-lapse video microscopy experiments used to investigate the effect of top collagen layer on the motility and spreading of hepatocytes showed that hepatocytes cultured on a single collagen layer exhibited flat, extended morphology and had lamellipodia extending out onto the collagen substrate (Ezzell et al., 1993). In contrast, hepatocytes in the sandwich gel maintained a polarized morphology for several weeks and showed minimal protrusive activity. Overlaying the second layer of collagen onto hepatocytes that had been cultured on a single layer for 4 days reversed cell spreading activity and stress fibers formation. "Sandwich rescue" also resulted in a decrease in actin and tubulin mRNA to levels similar to that found in new sandwich gel cultures and freshly isolated hepatocytes. These results show that the configuration of the external matrix has dynamic effect on cytoskeletal proteins in cultured rat hepatocytes and the addition of the top layer of collagen causes cultured hepatocytes to mimic the in vivo state both "inside and out".

Culture Media Optimization and the Role of Proline

While our initial studies on the collagen sandwich system demonstrated long-term hepatocellular function in vitro, the medium used in these experiments contained fetal bovine serum. In an effort to simplify and improve the culture technique, a preliminary investigation into the medium requirements of these cultures were undertaken (Lee et al., 1992; 1993). L-Proline was shown to be critical for both the approach to steady-state and maximal level of protein secretion when hepatocytes were cultured in serum free media. The response of hepatocytes to proline showed a dose dependent relationship with the maximal activity obtained at a proline concentration of 20 μg/ml.† The maximal rate of protein secretion in serum-free cultures was comparable to the estimated rate of protein secretion for intact liver as well as serum-containing cultures.

One possible expalnation the proline effect in serum-free cultures may involve the requirement of proline for newly synthesized extracellular matrix. Proline is one of the most abundant amino acids of collagen, and its derivative, hydroxyproline, plays an important role in the formation of the triple helical structure of collagen (Eyre, 1980; Kleinman et al., 1981). In sandwich cultures of hepatocytes, proline supplementation may stimulate native collagen synthesis by hepatocytes, and this newly formed collagen may be important for

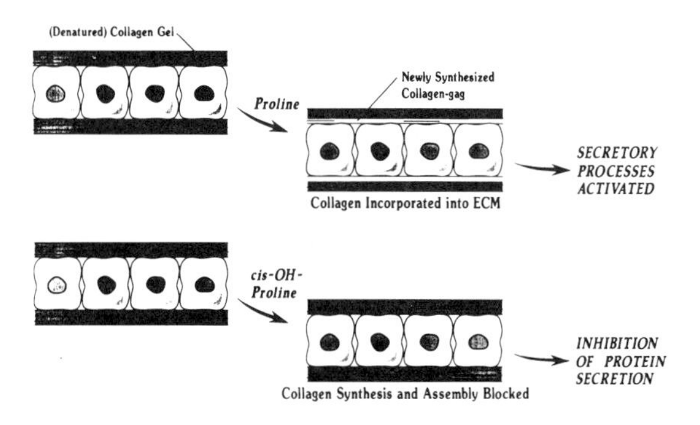

FIGURE 3. Hypothesis for proline effect on hepatocyte function.

† It is interesting to note that the amount of proline contained in the serum supplements is also about 20 μg/ml.

cell signaling (Figure 3). Two complementary lines of evidence support this hypothesis. First, hepatocyte cultures require proline continuously in order to sustain hepatocyte function. Second, hepatocyte function is sensitive to inhibition by *cis* -hydroxyproline throughout the culture period and this inhibition was completely reversible within 6-8 days which is precisely the length of time required for newly initiated cultures to attain maximal protein secretion levels. The sensitivity of the established cultures to *cis* - hydroxyproline suggests that the extracellular matrix may be undergoing continued remodeling in the sandwich culture. These results are consistent with our previously stated hypothesis that the top layer of collagen serves as a barrier function. The top gel layer can be seen to prevent the loss of endogeneously synthesized and newly secreted collagen and other extracellular matrix components into the surrounding medium and, thus, allows these components to coalesce around the hepatocyte.

Cryopreservation and Hypothermic Storage of Hepatocytes

The ultimate clinical success of any bioartificial liver device, whether extracorporeal or transplantable, relies on a readily available supply of large number of hepatocytes when the necessity for liver support arises. Because hepatocytes do not proliferate in culture, cryopreservation is critical for providing a stable supply of hepatocytes either from single or multiple sources so that sufficient numbers of cells may accumulate over time. In an attempt to design efficient protocols for cryopreservation of hepatocytes, we have investigated both the physicochemical aspects of cryopreservation and long-term hepatocyte function following well-defined freezing protocols. A specially designed freezing unit was used to provide controlled temperatures throughout a freeze-thaw cycle (Toner et al., 1993). Cooling rate, warming rate, and the final freezing temperature were evaluated as to their effect on hepatocyte function as judged by albumin secretion (Borel Rinkes et al., 1992b).

The major challenge to cells during freezing is not their ability to endure storage at very low temperatures, but rather, their ability to traverse through an intermediate temperature zone between -10 and -60°C during both cooling and warming. Intracellular ice formation can occur during rapid cooling as well as slow warming and result in irreversible cellular injury (Mazur, 1984). The actual cooling and warming rates yielding intracellular ice formation are a function of the cell type and depend on both the water permeability of the plasma membrane and the ice-nucleation parameters of the specific cell type under investigation (Toner et al., 1992). Our physicochemical and cryomicroscopy studies with sandwiched hepatocytes showed that hepatocytes cooled at rates faster than 10°C/min underwent intracellular ice formation (Hubel et al., 1991). Consistent with these cryomicroscopic observations, bulk freezing experiments with sandwiched hepatocytes showed a sharp decrease in survival for cooling rates faster than 10°C/min with a maximum at 5°C/min. Despite the use of what can be considered as an optimal cooling rate (i.e., the fastest cooling rate without lethal intracellular ice formation), long-term protein secretion was significantly reduced when final freezing temperatures were below -40°C and the warming rate was 5°C/min. Possible explanation for this observation might include the formation of innocuous very small ice crystals in the cell cytoplasm during cooling which may grow to larger sizes upon warming (i.e., recrystallization). Using rapid warming in the range of 400°C/min, the survival of sandwiched hepatocytes, as assessed by their long-term albumin secretion post-thawing, was increased to levels comparable (~ 75%) to those found in non-frozen controls (Borel Rinkes et al., 1992). These results show that rapid warming may, indeed, increase cell survival, most likely by not allowing sufficient time for recrystallization to take place. In summary, our results show that sandwiched hepatocytes can regain satisfactory levels of long-term differentiated function and structure following an optimized freeze-thaw protocol.

In addition to cryopreservation of hepatocytes for long-term storage, efficient short-term storage protocols would also serve as a useful purpose. Hypothermic cooling would provide a simple means to extend the shelf-life of hepatocytes in a bioartificial liver device so that cells could be preserved when the device is not in use.

We investigated effects of hypothermia at 4°C on albumin secretion from cultured hepatocytes in the absence of cryoprotective agents. Hepatocytes were exposed to hypothermic conditions (4°C) at the end of 8 days of culture in the sandwich configuration. Short exposure to 4°C (less than 4 hours) resulted in only a small decrease in albumin secretion rates following the hypothermic treatment (Figure 4). Longer exposures to 4°C resulted in greater impairment of albumin secretion rates (Figure 4). In addition, cultured hepatocytes were stained for F-actin before and after hypothermia to determine the effect of cold

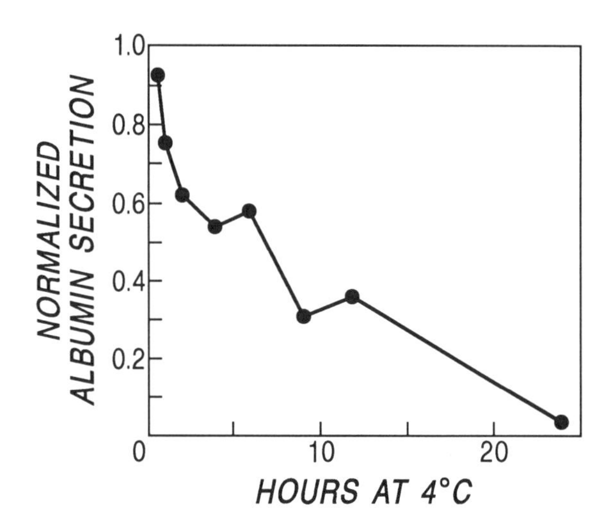

FIGURE 4. Normalized albumin secretion rate as a function of time at 4°C. Hepatocytes in the sandwich culture configuration were exposed to hypothermia after 8 days of culture to establish steady-state conditions. Normalized albumin secretion rate is defined as cumulative albumin secretion between days 11 and 14 of culture, normalized to the respective control value of each experiment. Data ponts represent the outcome of all experiments performed (mean ± SD; n = 2-6 per condition).

on cytoskeletal organization. Control hepatocytes had an actin distribution similar to in vivo cells and the actin was mostly membrane bound. On the other hand, following a 4 hour treatment, there was an increase in the cytoplasmic actin which showed a patchy morphology suggesting actin precipitation in the cytoplasm. However, these changes were mostly reversible after the normothermic conditions were reestablished. In contrast, the distribution of actin was drastically altered for those exposed to 4°C for 9 hours or more. Twenty-four hour after normothermia was re-established, all cells had lost their membrane bound actin and died. Our current studies are focused on extending these preliminary hypothermia studies to include studies in which cryoprotective agents such as polyethylene glycol are used to further extend the shelf-life of cultured hepatocytes.

IN VITRO RECONSTRUCTION OF TISSUE AND DESIGN OF DEVICES FOR LIVER SUPPORT

The clinical features of most forms of liver failure display a degeneration of all aspects of liver function. Thus, liver assist devices which utilize hepatocytes would be extremely versatile and afford hepatic support in a variety of clinical situations. Hepatocyte-based systems for liver replacement can be either transplantable or extracorporeal. Although there is considerable interest in hepatocyte transplantation, this area is clearly in its infancy (Cima et al., 1991). Extracorporeal systems, also called bioreactors, involve an external circuit that treats blood or plasma from the patient (Arnaout et al., 1990; Sussman et al., 1992; Shnyra et al., 1990; Uchino et al., 1988). Although bioreactors have been used in animals and isolated clinical cases, many important issues remain yet to be resolved. For example, it is not clear how long liver specific function will last in a device, and the effect of oxygen and nutrient limitations on the loss of initial cell mass has yet to be quantitatively evaluated. Also, a clear demonstration, utilizing appropriate control groups, of improved survival due to the extracorporeal device remains to be achieved. All of these issues will be addressed in time provided that careful study follows the instances where clinical benefit is shown.

Effect of Oxygen on Hepatocyte Attachment and Spreading.

Regardless of the device configuration contemplated, hepatocytes must be firmly attached and spread on a surface in order to function. Since any bioreactor would contain a large number of hepatocytes in a small volume and since the spreading of cells is an energy requiring process, oxygen-limiting conditions may be experienced during the seeding of the bioreactor with hepatocytes. We have recently begun to study the role of oxygen by examining the kinetics of attachment and spreading of hepatocytes onto a flat collagen gel under different oxygen tensions (Rotem et al., 1993). The effect of oxygen on the attachment of hepatocytes was determined under increasing gas phase oxygen tensions after 1 hour of seeding onto the collagen gel. Increasing the oxygen tension from 0 to 10% caused an increase of the cell attachment from ~ 40% to ~ 100%. Further increase in oxygen concentration did not have any substantial effect. Furthermore, this increase in the gas phase oxygen tension did also increase the extent of hepatocyte spreading. These results demonstrate that the anchorage-dependent hepatocytes require a critical amount of oxygen to attach and spread in order to function optimally.

Bioreactor Configurations

Two of the main tasks facing the bioengineer working on the design of a bioartificial liver support device are: (1) to be able to accommodate a large number of metabolically active cells with an adequate nutrient supply and waste removal system, and (2) to maintain these cells in differentiated state. With these two tasks in mind, we have begun investigating a number of three-dimensional designs which incorporate our stable sandwich culture method. Three different approaches will be discussed: (1) microcarrier based systems, (2) micropatterning devices, and (3) porous polymeric matrices.

Although microcarriers provide a large surface area in a relatively small volume, their spherical shape makes cell attachment less straightforward. When attaching hepatocytes to microcarriers, we must maximize the number of cells per microcarrier because hepatocytes do not divide to any appreciable extent. This differs from most microcarrier applications where one attaches a small number of proliferating cells to each microcarrier. As an extension to our previously described hepatocyte attachment studies, we have recently investigated the attachment of primary rat hepatocytes onto Cytodex 3 microcarriers (Foy et al., 1993). The maximum number of hepatocytes per microcarrier obtained was approximately 70 to 100 when gas phase oxygen tensions of 50% and above were used. This number was drastically reduced to less than 20 hepatocytes per microcarrier for oxygen tensions lower than 21%. Figure 5 shows phase contrast photomicrographs of microcarriers after 8 hours in either 70% or 21% gas phase oxygen tension. The maximum number of hepatocytes which can be attached to a microcarrier can be calculated on the basis of the surface area of an average microcarrier and the projected surface area of an average hepatocyte. The projected surface area for freshly isolated hepatocytes is about 375 μm^2, while the projected surface area for hepatocytes after a few days in culture on flat surfaces is about 1250 μm^2. About 275 freshly isolated cells or about 75 spread hepatocytes will fit on a microcarrier of mean diameter 175 μm. Thus, our experimental results for the maximal number correspond well to those obtained by purely theoretical concerns. We are currently investigating several approaches for incorporating the sandwich culture geometry within a microcarrier-based reactor.

A second reactor design relies on the efficient micropatterning of collagen-treated surfaces interspersed between non-charged hydrophilic regions which are devoid of cells. Given a flat geometry the sandwich configuration can be established by entrapping the hepatocytes between two micropatterned surfaces. In this type of configuration, the regions devoid of cells would approximate the upper and lower walls of the hepatic sinusoids. Each hepatocyte would be exposed to perfusate from both sides while bounded by collagen from top and bottom. This geometry can be ultimately scaled up with multiple units of sandwiched micropatterned surfaces. Figure 6 shows some preliminary results of selective adhesion of cells to a micropatterned substrate. As can be seen ,hepatocytes attached only to the collagen treated surfaces; whereas, the non-attaching regions are substantially devoid of cells.

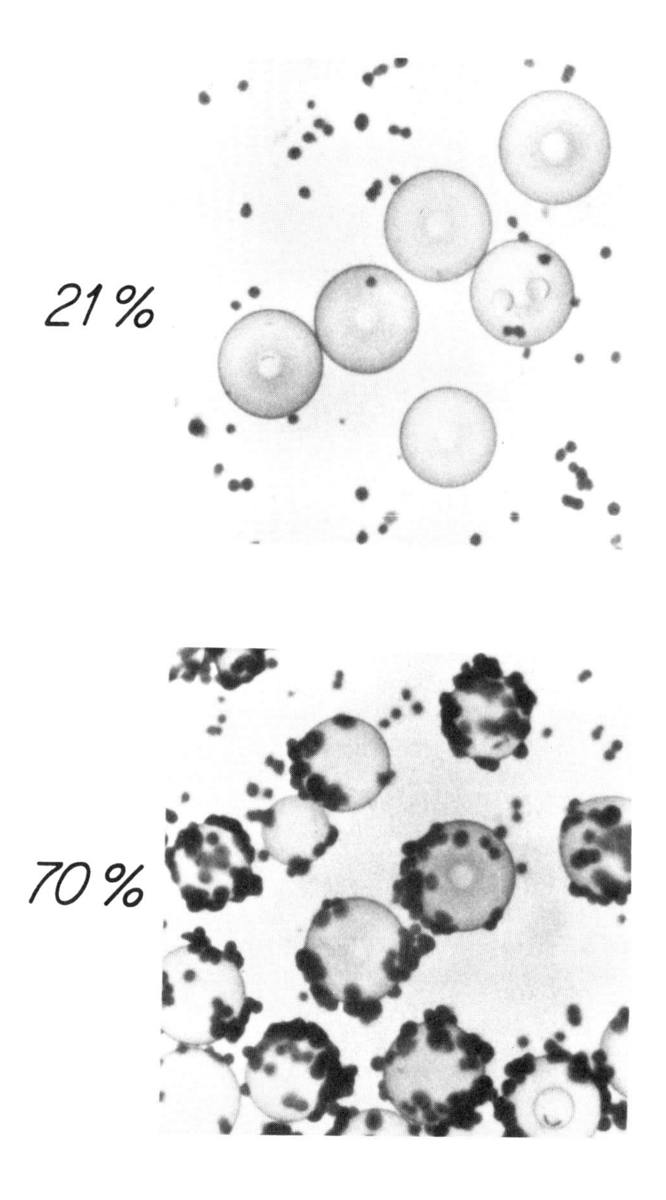

FIGURE 5. Phase contrast photographs of microcarriers after 8 hours incubations at either (A) 70% gas phase oxygen tension, or (B) 21% gas phase oxygen tension.

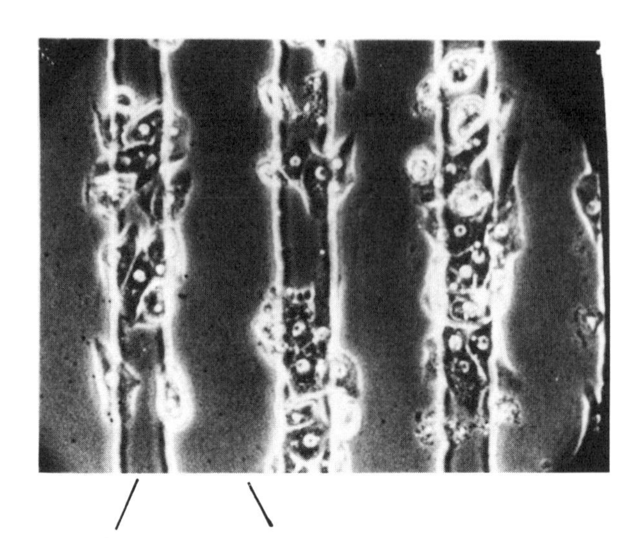

FIGURE 6. Phase contrast photograph of hepatocytes on a micropatterned surface. The width of the attaching layer with hepatocytes is about 50 μm and the width of the non-attaching surface is about 100 μm. Micropatterned surface was coated with dilute collagen gel by dipping the surface into liquid DMEM+collagen at 4°C. Following diping, liquid DMEM+collagen remained only on the hydrophilic surfaces. The thin layer of collagen was then gelled by exposure to 37°C prior to the seeding with hepatocytes.

In a third approach, we are investigating the possibility of creating a microporous hollow fiber consisting of pore walls of 100 μm in diameter. These microporous hollow fibers can be processed by solidification techniques similar to the one developed for nerve regeneration studies (Loree, 1988). However, instead of creating cylinders consisted of longitudinal pores, it is our hope to make hollow fibers with pores distributed radially across their walls. In this way, one could hypothetically seed the biomaterial with hepatocytes and then introduce endothelial cells to establish the sandwich geometry.

In summary, hepatic tissue engineering brings together fundamental engineering and biological sciences with the ultimate goal of creating a liver support device. Although several unresolved issues critical to the clinical use of bioartificial liver devices remain, these problems should be solvable using tissue engineering approaches.

ACKNOWLEDGEMENTS

The research from our laboratory reviewed here was supported by grants from the National Institutes of Health (DK-43371, DK-41709, GM-21700, and GM-07035), the Shriners Hospitals for Crippled Children, and The Whitaker Foundation.

REFERENCES

Arnaout WS, Moscioni AD, Barbour RL, and Demetriou AA. (1990): Development of bioartificial liver: bilirubin conjugation in Gunn rats. *J. Surg. Res.* 48: 379-382.

Bader A, Borel Rinkes IHM, Closs E, Ryan C, Toner M, Cunningham J, Tompkins RG, and Yarmush ML (1992): A stable long-term hepatocyte culture system for studies of physiologic processes: Cytokine stimulation of the acute phase response in rat and human hepatocytes. *Biotech Prog* 8: 219-225.

Bell E, Rosenberg M, Kemp P, Gay R, Green GD, Muthukumaran N, and Nolte C (1991): Recepies for reconstituting skin. *J. Biomech. Eng.* 113: 113-119.

Berry MN, and Friends DS (1986): High-yiled preparation of isolated rat liver parenchymal cells. *J. Cell Biol.* 43: 506-520.

Borel Rinkes IHM, Toner M, Ezzell RM, Tompkins RG, and Yarmush ML (1992a): Effect of dimethyl sulfoxide on cultured rat hepatocytes in sandwich configuration. *Cryobiol* 29: 443-453.

Borel Rinkes IHM, Toner M, Tompkins RG, and Yarmush ML (1992b): Long-term functional recovery of hepatocytes after cryopreservation in a three-dimensional culture configuration. *Cell Transplantation* 1: 281-292.

Cima LG, Vacanti JP, Vacanti C, Ingber D, Mooney D, and Langer R (1991): Tissue engineering by cell transplantation using degradble polymer substrates. *J. Biomech. Eng.* 113: 143-151.

Colton CK, and Avgoustiniatos ES (1991): Bioengineering in development of the hybrid artificial pancreas. *J. Biomech. Eng.* 113: 151-170.

Dunn JCY, Yarmush ML, Tompkins RG, and Koebe H (1989): Hepatocyte function and extracellular matrix geometry: Long-term culture in a sandwich configuration. *FASEB J* 3: 174-177.

Dunn JCY (1991): Long-term culture of differentiated hepatocytes in a collagen sandwich configuration. Ph.D. Thesis, Department of Chemical Engineering, Massachusetts Institute of Technology, Cambridge, MA.

Dunn JCY, Tompkins, RG, and Yarmush, MY (1991): Long-term in vitro function of adult hepatocytes in a collagen sandwich configuration. *Biotech Prog* 7: 237-245.

Dunn J, Tompkins RG, and Yarmush ML (1992a): Hepatocytes in collagen sandwich configuration: Evidence for transcriptional and translational control. *J Cell Biol* 116: 1043-1053.

Dunn JCY, Friedberg JS, Tompkins RG, and Yarmush ML (1992b): Hepatocytes from rat liver perfusions: physicochemical effects on polyribosome size. *ASAIO J.* 38: 841-845.

Dunn JCY, Yarmush MLY, and Tompkins RGT (1993, in press): Dynamics of transcriptional and translational processes in hepatocytes cultured in collagen sandwich. *Biotech Bioeng* .

Eyre DR (1980): Collagen: molecular diversity in the body's protein scaffold. *Science* 207: 1315-1322.

Ezzell RM, Toner M, Hendricks K, Dunn JCY, Tompkins RG, and Yarmush ML (in preparation): Dynamics and structure of cytoskeletal organization of cultured hepatocytes: effects of the collagen gel.

Foy, BD, Lee, J, Morgan, J, Toner, M, Tompkins, RG, and Yarmush, ML (1993, in press): Optimization of hepatocyte attachment to microcarriers: importance of oxygen. *Biotech and Bioeng.*

Harris CL, Toner M, Hubel A, Cravalho EG, Yarmush ML, and Tompkins RG (1991): Cryopreservation of isolated hepatocytes: intracellular ice formation under various physical and chemical conditions. *Cryobiol* 28: 17-30.

Hubel A, Toner M, Cravalho EV, Yarmush ML, and Tompkins RG (1991): Intracellular ice formation during the freezing of hepatocyte culture in a double collagen gel. *Biotech Prog* 7: 554-559.

Kleinman HK, Klebe RJ, and Martin GR (1981): Role of collagenous matrices in the adhesion and growth of cells. *J. Cell Biol.* 88: 473-485.

Knowles BB, Howe CC, and Aden DP (1980): Human hepatocellular carcinoma cell lines secrete the major plasma proteins and hepatitis B surface antigens. *Science* 209: 497-499.

Koebe H, Dunn JCY, Toner M, Sterling LM, Hubel A, Cravalho EG, Yarmush ML, and Tompkins RG (1990): A new approach to the cryopreservation of hepatocytes in a sandwich configuration. *Cryobiol* 27: 587-594.

Kreamer BL, Staecker JL, Sawada GL, Sattler MT, Hsia MT, and Pitot H (1986): Use of a low-speed, iso-density percoll centrifugation method to increase viability of isolated rat hepatocytes preparations. *In Vitro Cell. Dev. Biol.* 22: 201-211.

Lee J, Morgan JR, Tompkins RG, and Yarmush ML (1992): The importance of proline on long-term hepatocyte function in a collagen sandwich configuration: regulation of protein secretion. *Biotech and Bioeng* 40: 298-305.

Lee J, Morgan JR, Tompkins RG, and Yarmush ML (1993, in press): Proline mediated enhancement of hepatocyte function in a collagen gel sandwich culture configuration. *FASEB J.*

Loree, HM (1988): A freeze-drying process for fabrication of polymeric bridges for peripheral nerve regeneration. M.S. degree thesis, Department of Mechanical Engineering, Massachusetts Institute of Technology, Cambridge, MA.

Mazur P (1984): Freezing of living cells: mechanisms and implications. *Am. J. Physiol.* 143: C125-C142.

Reese JA, and Byard JL (1981): Isolation and culture of adult hepatocytes from liver biopsies. *In Vitro* 22: 935-940.

Rotem A, Toner M, Tompkins RG, and Yarmush ML (1992): Oxygen uptake rates in cultured rat hepatocytes. *Biotech Bioeng* 8: 227-232.

Rotem A, Toner M, Bhatia S, Foy, BD, Tompkins RG, and Yarmush ML (in preparation): The effect of oxygen on attachment and spreading of cultured rat hepatocytes.

Ryan C, Carter EA, Jenkins R, Sterling L, Yarmush ML, Malt R, and Tompkins RG (1993): Isolation and long-term culture of human hepatocytes. *Surgery* 113: 48-54.

Seglen PO (1976): Preparation of isolated rat liver cells. *Methods Cell Biol.* 13: 29-83.
Shnyra A, Bocharov A, Bochkova N, and Spirov V (1990): Large-scale production and cultivation of hepatocytes on Biosilon microcarriers. *Artif. Organs* 14: 421-428.
Sussman NL, Chong MG, Koussayer T, He D, Shang TA, Whisennad HH, and Kelly JH (1992): Reversal of fulminant hepatic failure using an extracorporeal liver assist device. *Hepatology* 16: 60-65.
Tompkins RG, Carter EA, Carlson J, and Yarmush ML (1988): Enzymatic function of alginate immobilized rat hepatocytes. *Biotech Bioeng* 31: 11-18.
Toner M, Borel Rinkes IHM, Cravalho EG, Tompkins RG, and Yarmush ML (1993, in press): A controlled rate freezing device for cryopreservation of biological tissue. *Cryo-Lett.*
Toner M, Cravalho EG, Tompkins RG, and Yarmush ML (1992): Transport phenomena during freezing of isolated hepatocytes. *AIChE J* 38: 1512-1522.
Uchino J, Tsuburaya T, Kumagai F, Hase T, Hamada T, Komai T, Funatsu A, Hashimura E, Nakamura K, and Kon T (1988): A hybrid bioartificial liver composed of multiplated hepatocyte monolayers. *ASAIO Trans.* 34: 972-977.
Yarmush ML, Dunn J, and Tompkins RG (1992a): Assessment of artificial liver support technology. *Cell Transplantation* 1: 410-423.
Yarmush ML,Toner M, Dunn JCY, Rotem A, Hubel A, and Tompkins RG (1992b): Hepatic tissue engineering: Development of critical technologies. *Ann N.Y. Acad Sci* 665: 472-485.

Section V

Physical Forces as Requirements for Gene Expression, Growth, Morphogenesis, and Differentiation

EVIDENCE FOR THE ROLE OF PHYSICAL FORCES IN GROWTH, MORPHOGENESIS AND DIFFERENTIATION

Richard Skalak
Department of Applied Mechanics and Engineering Sciences, Bioengineering
University of California, San Diego

INTRODUCTION

Historically, evidence for the role of physical forces and internal stresses on growth, morphogenesis and differentiation has been observed and utilized for a long time. Both soft and hard tissues need stress for their proper development, maintenance, and repair. However, the detailed mechanisms of cellular response to physical forces are not known at a molecular level. The sensing, transduction, and monitoring mechanisms associated with stress effects on cells are just now being elaborated.

The purposes of the present article are to provide an overview and some background of the role of stress in growth and repair of tissues and to relate this to tissue engineering in general and the present symposium in particular.

HARD TISSUES

It happens that the present symposium articles following are devoted primarily to soft tissues, but historically attention has been focused earlier and more attention has been given to the role of stress on growth and repair in bone than in soft tissues. For the growth of bone Wolff's law is often quoted to indicate that bone particularly responds to stress ,or the lack thereof, to increase or decrease its strength or to change its form (Wolff, 1892). While there is no doubt about the basic fact that bone grows and remodels in response to stress distribution, the cellular mechanisms of sensing stress and biochemical pathways of response are still not fully known and constitute an area of active research.

Attempts have been made to quantify Wolff's law. For example, Cowin (1989) has interpreted Wolff's law as the coincidence of principal axes of stress with the principal axes of trabecular architecture and developed the fabric tensor which describes this situation geometrically. Such macroscopic theories are able to correlate the observed architecture and applied stress.

The initial growth and development of bone has also been interpreted in terms of response to stress (Carter, 1989);. The bone growth, and maturation of bone is described as proliferation, degeneration and ossification of cartilage. The resulting tissue depends on the stress field applied both in the type of tissue produced and its orientation and strength. The theory is successful in predicting the morphology of long bones from embryonic cartilage anlagen to maturity and senasence.

The experience of astronauts in zero gravity environment supplies evidence that stress is necessary to the maintenance of normal bone density. Without stress, bone loses some of its density and strength.

Tendons also respond to increased stress by increasing their strength, although this is a small effect (Woo, 1989).

Thus, there is a variety of evidence for the regulating role of stress in growth, morphogenesis and differentiation of hard tissues. But in regard to tissue engineering this knowledge has not led to development of any living replacement for bone. This is due in part to the fact that any living replacement will require a blood circulation to maintain viability. Provision of a living replacement with a functional circulation is not feasible at present. Even in autologous bone grafts, the living cells die and are replaced in time by infiltration from surrounding bone and a new circulation is concurrently established.

SOFT TISSUES

The current research on the role of mechanical stress on the growth of soft tissue is well-represented in the present symposium. The following papers demonstrate that stress affects the growth of endothelial cells, vascular walls, cartilage, cardiac myocytes and cell differentiation. Further, other chapters of the present volume will describe practical progress in the development of living replacements for skin and for blood vessels. These two areas are probably the most advanced and successful applications that may be described as tissue engineering. (See Hansbrough, 1992 and Greisler, 1992).

The current situation in regard to use of stress in tissue engineering to mold soft tissue replacement is similar to that in bone. The effects of stress are clear, but so far, no clinically useful procedures have been developed that utilize stress effects in developing s soft tissue replacement. This is a potential that is not yet capitalized. Stress may be regarded as an analog to the steering mechanism of an automobile. The cellular machinery and biochemistry of cell division and growth are analogous to the engine and gasoline of the automobile, but the appropriate use of stress as a steering mechanism may allow reaching objectives that are not accessible by simply driving straight ahead.

CONCLUSION

The influential role of mechanical stress in growth, morphogenesis and differentiation has been amply demonstrated in both hard and soft tissues. It may be a more meaningful question to ask whether there is any tissue whose growth is not affected by stress. However, to date, the knowledge of the effects of stress on growth have not been incorporated in clinical practice of tissue engineering.

REFERENCES

Carter DF (1989) The Control of Skeletal Biology by Mechanical Energy. In: Fung et al. (Eds), *Progress and New Directions of Biomechanics*. Mita Press, Tokyo, 1989, pp. 13-22.

Cowin SC (1989): *A Geometric Analog Model of Wolff's Law of Trabecular Architecture*. In: Fung et al. (Eds). *Progress and New Directions of Biomechanics*, Mita Press, Tokyo, pp. 333-342.

Fox CF (Ed.) (1990): *Tissue Engineering*. Supplement Volume 14E, *J Cellular Biochemistry*.

Fung YC, Hayashi K and Seguchi Y (Eds.) (1989): *Progress and New Directions of Biomechanics*. Mita Press, Tokyo.

Greisler R (1992): *New Biologic and Synthetic Vascular Prostheses*, CRC Press, Boca Raton, Florida.

Hansbrough JF (1992): *Wound Coverage with Biologic Dressings and Cultured Skin Substitutes*. R. G. Landes Co., Boca Raton, Florida.

Skalak R and Fox CF (Eds.) (1988): *Tissue Engineering*. Alan R. Liss, Inc., New York, 343 pp.

Skalak R and Heineken FG (Eds.) (May 1991): *Tissue Engineering: A Brief Overview*. Special Issue of the J. Biomechanical Engineering, American Society of Mechanical Engineers, New York. Vol. 113.

Wolff J (1986) *Das Gesetz der Transformation der Knochen*. (*The Law of Bone Remodeling*, translated by P. Maquet and R. Furlong, 1986), Springer-Verlag, Berlin, 1892.

Woo SL and Seguchi Y (1989): *Tissue Engineering - 1989* Bioengineering Division Vol. 14, American Society of Mechanical Engineers, New York, 145 pp.

PHYSICAL STRESS AS A FACTOR IN TISSUE GROWTH AND REMODELING

Yuan-Cheng Fung and
Shu-Qian Liu
Institute for Biomedical Engineering
University of California, San Diego
La Jolla, CA 92093-0412

INTRODUCTION

The science of tissue engineering seeks to understand the behavior of living tissues, to control their growth and resorption, to produce them in desired quantities and to make them work for the benefit of man. It rests on several foundations, of which the most important are the molecular biological one and the biomechanical one. The importance of physical stress is well known. For example, high blood pressure or excessive flow causes hypertrophy of the heart.; whether a healing fracture in bone will calcify or not depends on the nature and magnitude of the stress acting in the bone. Similarly, the success of culturing and application of artificial tissues to man will depend on proper stress conditions.

The difficulty of tissue engineering science is that the living tissue is extremely complex and involves too many variables. To cut through the maze we advocate the use of engineering methods. Two principles of engineering investigations are: 1) Vary one variable at a time if possible. 2) Study the response to the simplest input function. The simplest input is the step function if possible. Response to a step function is called the *indicial response*. If the system were linear, than with a convolution of *indicial response* one can compute the response to any arbitrary input function. If the system were nonlinear, then we study the deviations from linearity. In this article we illustrate these principles by examples relevant to the tissue engineering of blood vessels.

INDICIAL RESPONSE OF A BLOOD VESSEL TO BLOOD PRESSURE.

The question is to study the effects of stress on blood vessel wall, and separate the influence of the shear stress due to blood flow and influence of the tensile stress due to blood pressure. Fortunately, it is possible to vary pressure and flow separately, one at a time, in the lung. If we breath a gas whose oxygen content is lower than normal, then the smooth muscle cells in the pulmonary arteries will contract while the cardiac output remain unchanged. This will cause an increase of pulmonary blood pressure while the pulmonary blood flow is almost unchanged. When a rat is put into a hypoxic chamber with 10% O_2 and 90% N_2 at the atmospheric

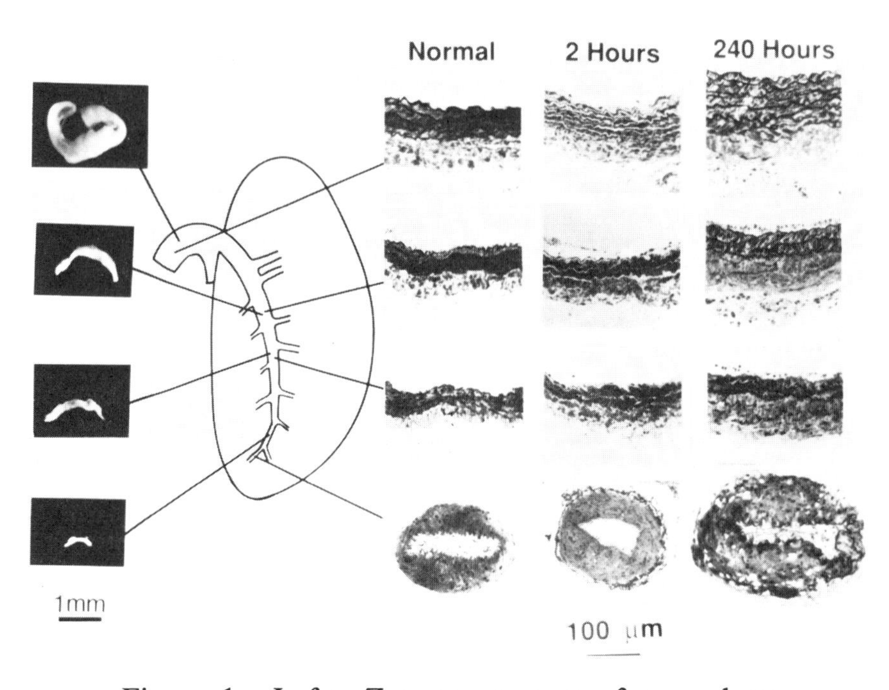

Figure 1. Left: Zero-stress state of normal rat pulmonary artery. Endothelium downward. Right: Histology of normal and hypertensive rats. Collated from Y. C. Fung and S.Q. Liu J. Appl. Physiol. 70:2455-2470, 1991.

pressure, the pulmonary arterial blood pressure will rise 7 mm Hg within a minute, remain at this level for a week, then rise further slowly for another 4 mm Hg in 30 days. The pressures in the pulmonary vein at the left atrium and in the aorta will remain practically unchanged in this period. Hence the blood pressure in pulmonary artery rises as a step function while the flow is constant; and the tensile stress in the artery rises but the shear does not.

The consequence of this change is shown in Figure 1. At the *center* of Fig. 1 is a sketch of the main pulmonary artery of the rat. The photographs at the *right* show the structure of the vessel wall. Clear changes can be seen at two hours of high blood pressure and progressively later. At two hours, the endothelial cells are swollen, the intima becomes edematous, and the elastin in the internal elastic lamina of the intima and the elastic lamina of the media has a change of stain color. The intimal and medical layers thicken rapidly at first in the period from 48 to 240 hours of high blood pressure, and then slowly from 240 hrs to 720 hrs. The outer collagenous layer of the adventitia thickens later. Thus the remodeling progressed nonuniformly in the arterial wall.

The photographs shown at the *left* side of Fig 1 are the cross sections of the normal pulmonary artery at the sites indicated. We first cut the artery transversely into rings; then cut radially on the side facing the heart. The rings spring open into sectors. The angle subtended by radii drawn from the midpoint of the inner wall to the tips of the sector is the *opening angle*. The opening angle of normal pulmonary artery is seen to be about 360° at the main trunk, and smaller at other locations; and can change another 100° or more in the course of tissue remodeling (see Ref. 17).

The shape of the sectors shown on the left side of Fig. 1 is the zero-stress state of the pulmonary artery. In the study of tissue remodeling it is best to compare tissue morphology at the zero-stress state, because at this state the shape is not distorted by stresses.

We studied further details of the tissue remodeling of arteries, including the indicial responses of the amounts of collagen fiber bundles, elastin laminas, fibroblasts, and the shape, size, and proliferation of the smooth muscle cells (Refs 15-23). We measured not only the morphological changes, but also the changes of mechanical properties of the vessel wall as a function of time following the sudden onset of hypertension (Ref. 19) or diabetes (Ref. 23).

These indicial responses can be stated in mathematical form with a number of material constants determined by experiments. Once obtained, then we can use them to analyze complex practical problems, making predictions, and prepare for validation clinically.

Other authors (Refs 1-14) have studied the morphology, cell proliferation, DNA, gene expression, extracellular matrix, and other features. Omens and Fung (24) extended the study to the left ventricle. We studied also the aorta (17), microvessels (21), veins (25), and trachea (26).

GUIDANCE FROM IRREVERSIBLE THERMODYNAMICS

From the point of view of thermodynamics, tissue growth is an irreversible process. In an irreversible process. there is entropy production within the system. According to the second law of thermodynamics, entropy production is non-negative, and can be written as a sum of the products of "generalized fluxes" and their conjugate "generalized forces." The phenomenological laws are those that relate the generalized fluxes to the generalized forces, or vice versa. Famous physicists like Lord Kelvin, Onsager, and Prigogine, engineers like Darcy, and physiologists like Starling, have studied phenomenological laws. In biomechanics, constitutive equations are often based on the identification of fluxes and forces in entropy production, (Ref. 15, 16).

In a growing tissue, the heat change is the sum of the changes in internal energy, the product of chemical potential and the rate of increase of mass, the product of muscle force and the rate of contraction, the product of stress and the rate of change of strain, etc. If we take as generalized fluxes in the entropy production the strain rate, the rate of contraction of muscle, the rate of change of the mass per unit volume of certain materials (e.g., calcium in bone, smooth muscle in blood vessel, collagen in vessel wall), then the conjugate generalized forces are the stresses, muscle tension, and chemical potentials divided by the absolute temperature.

Thus we expect a phenomenological law which relates the rate of increase of the mass of a material in a tissue with the stress, muscle tension, and gradients of chemical potentials divided by the absolute temperature. Other things being equal, then there is a stress-growth law.

A stress-growth law proposed by Fung (1990) is shown in Fig. 2. The horizontal axis is stress. The vertical avis is the rate of increase of mass per unit volume of the specific material of interest. A positive $\dot{m}$ means growth. A negative $\dot{m}$ means resorption. The points *a b, c* represent states of equilibrium at normal living condition. In the neighborhood of the *a*, an increase of stress correlates with growth, a decrease of stress correlates with resorption. At the points *b* and *c* the reverse in true. At zero stress the rate $\dot{m}$ is positive if tissue culture is possible. For bone the condition at *a* is known as Wolff's law (Ref. 16). For blood vessels, heart, and lung, the evidences reviewed above supports the proposed law in the neighborhood of *a*. The rest remains to be verified.

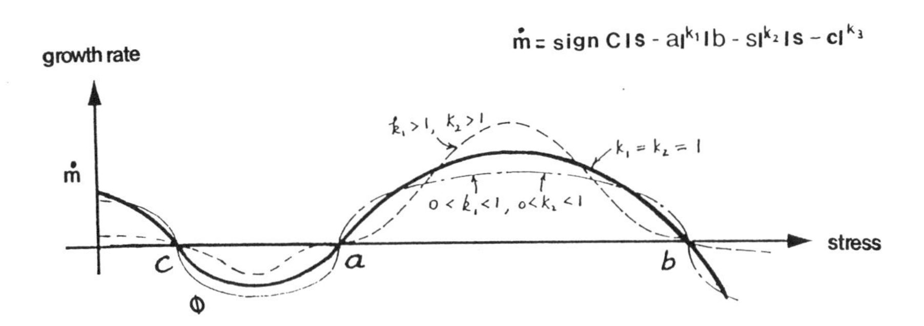

Figure 2. A proposed stress-growth law, Fung (1990). Sign means the ± sign of (s-a) (b-s) (s-c).

CONCLUSION

We propose to use engineering method to obtain the indicial responses to step changes in tensile stress and shear stress in soft tissues, and to determine a stress-growth law, the mathematical form of which is given in Figure 2, where s is stress, and a,b,c,C, k_1, k_2, k_3 are constants which depends on biochemical factors. A quantitative determination of the stress-growth law is the objective of biomechanical research in tissue engineering.

ACKNOWLEDGEMENT

The support of NSF through Grant No. NSF BCS 89-17576 and NIH through grants 2R01 HL 26647 and PPG HL 43026 is gratefully acknowledged.

REFERENCES

1) Atkinson, J.E. et al, (1987) *J. Appl. Physiol.* 62:1562-1568.
2) Hung, K.S. et al. (1986) *Acta Anat.* 126:13-20.
3) Kerr, I.S. et al (1984) *J. Appl. Physiol.* 57:1760-1766.
4) McKenzie, J. C. et al. (1984) *Blood Vessels* 21: 80-89.
5) Mecham, R.P. et al. (1987) *Science*, 237: 423-426.
6) Meyrick, B. et al (1978) *J. Anat.* 125: 209-221.
7) Meyrick, B. and Reid, L. (1978) *Lab. Invest.* 38:188-200. (1979) *Am. J. Pathol.* 96: 51-70. (1980) *Lab. Invest.* 42: 603-615 (1980) *Am. J. Pathol.* 100:151-178.
8) Poiani, G. J. et al (1990) *Circ Res.* 66:918-978.
9) Rabinovitch, M. et al (1979) *Am. J. Physiol* 236:H818-H827.
10) Reid, L. *Chest,* (1986) 89:279-288.
11) Smith, P. et al (1974), *J. Pathol.* 112: 11-18.
12) Sobin, S.S. et al. (1983), *J. Appl. Physiol.* 55: 1445-1455.
13) Stenmark, K.R. et al. (1987) *J. Appl. Physiol.* 62:821-830.
14) Thompson, B.T. et al (1989) *J. Appl. Physiol* 66: 920-928.
15) Fung, Y.C. (1990) *Biomechanics: Motion, Flow, Stress and Growth,* New York, Springer-Verlag.
16) Fung, Y.C. (1993). *Biomechanics: Mechanical Properties of Living Tissues.* 2nd ed., Springer Verlag, N.Y.
17) Fung, Y.C. and Liu, S.Q., (1989)*Circ. Res.* 65: 1340-1349.
18) Fung, Y.C. and Liu, S.Q., (1991)*J. Appl. Physiol.* 70: 2455-2470.
19) Fung, Y.C. (1991)*Ann. Biomed. Eng.* 19: 237-249
20) Fung, Y.C. and Liu, S.Q., (1992).*Am. J. Physiol.* 262: H544-H552
21) Liu, S.Q. and Fung, Y.C. (1988).*J. Biomech. Eng.* 110: 82-84.
22) Liu, S.Q. and Fung, Y.C. (1989) *J. Biomech. Eng.* 111:325-335.
23) Liu, S.Q. and Fung, Y.C. (1992) *Diabetes*, 41: 136-146
24) Omens, J.H. and Fung, Y.C. (1990) *Cir. Res.* 66: 37-45.
25) Xie, J.P. et al, (1991) *J. Biomech. Eng.* 113: 36-41.
26) Han, H.C. and Fung, Y.C. (1991) *J. Biomech.* 24: 307-315.

MECHANICAL STRESS EFFECTS ON VASCULAR ENDOTHELIAL CELL GROWTH

Robert M. Nerem, Masako Mitsumata, and Thierry Ziegler
Biomechanics Laboratory and School of Mechanical Engineering
Georgia Institute of Technology

Bradford C. Berk and R. Wayne Alexander
Division of Cardiology, Emory University School of Medicine

INTRODUCTION

Surgery is extensively used in the treatment of vascular disease as a procedure for restoring blood flow to ischemic areas. An example of this is coronary bypass surgery which has become widely used as a treatment for heart disease, with the surgical implantation of multiple bypasses being quite common. In this procedure native vessels, either the mammary artery or the saphenous vein, are generally employed as the vascular graft, i.e. the substitute blood vessel . The use of artificial blood vessels made of synthetic biomaterials also has been investigated.

More recently, researchers have turned their attention to the use of tissue engineering in the development of a biological substitute blood vessel. Both acellular and cellular approaches are being employed. In the latter case there again are two approaches, with one being the endothelial seeding of synthetic biomaterials (Zilla et al., 1987) and the other being the reconstitution of a blood vessel out of vascular endothelial cells (EC), smooth muscle cells (SMC), and extracellular matrix (Jones, 1982; Weinberg and Bell, 1986). Whatever the case, it is critical to understand those factors which regulate vascular growth.

In vivo vascular EC reside in a mechanical environment imposed by the dynamics of blood flow. As the interface between blood and the vessel wall, associated with the flow of blood through an artery the endothelium is exposed to a tangential force per unit area called shear stress and a normal stress called pressure. The endothelial cell both "sees" pressure directly and "rides" on a basement membrane which is undergoing cyclic stretch. This mechanical environment is quite complex; however, if we are to understand those factors which regulate vascular EC growth, it is important that we understand the influence of this hemodynamically-imposed environment.

There is some *in vivo* evidence suggesting an influence of flow on vascular EC proliferation (Wright, 1968; Caplan and Schwartz, 1973; Schwartz et al., 1983). Schwartz and his co-workers report a higher cell turnover rate in regions characterized by polygonally-shaped EC than that found in regions of more elongated EC. From other *in vivo* studies (Levesque et al., 1986; Kim et al., 1989) and from cell culture studies (Nerem and Girard, 1990), it has been demonstrated that EC are more elongated under the influence of high shear and rather polygonal under conditions of no flow or low shear. The *in vivo* evidence thus is suggestive of a flow inhibition of cell proliferation.

As important as *in vivo* studies are, it is difficult to quantify the details of the local vessel wall flow environment. Thus, a number of laboratories have turned their attention to the use of cell culture to study the influence of flow on vascular endothelial biology. Although not necessarily a physiologic model, cell culture provides the opportunity to study the response of vascular EC to a well defined flow environment. Over the past decade a wide variety of studies have been conducted in a number of different laboratories (Nerem and Girard, 1990). A major focus of our group has been on both the intrinsic and extrinsic growth programs of vascular EC, and it is results from these studies which are presented here.

CELL PROLIFERATION

In our laboratory the influence of flow on cell replication has been investigated through the measurement of cell density as a function of time in subconfluent bovine aortic endothelial cell (BAEC) monolayers grown on a Mylar surface under conditions of laminar, steady state flow. These data demonstrate that the rate of increase of cell density with time is decreased under steady flow conditions as compared to a static, no flow, control culture (Levesque et al., 1990). Exposure of EC to simple, non-reversing pulsatile flow (1 Hz, sinusoidal waveform) even more dramatically inhibited EC proliferation, with the decrease in cell growth being 50 percent greater than that in response to steady-state flow.

To demonstrate that this flow-related decrease was due to an influence on cell cycle, ^{3}H-thymidine autoradiography was used to compare the rate of DNA synthesis in EC immediately after pre-conditioning with flow for 24 hours to that of EC in static culture for 24 hours. A clear decrease in the rate of ^{3}H-thymidine uptake was apparent in EC which had been preconditioned by flow. This result is supported by flow cytometry measurements which indicate that EC populations exposed to flow exhibit a dramatic decrease in the percentage of cells in S phase as compared to control, static culture populations (Mitsumata et al., 1991a). Concomitant with this was a commensurate increase in the percentage of cells in G_0/G_1 for monolayers exposed to flow, thus suggesting that the influence of flow is to inhibit entry into S phase.

There of course is the possibility that there might be increased detachment of cells dividing in the presence of flow. This idea is based on the classical picture of a cell rounding up in the process of undergoing division and thus offering a large frontal area over which a flow-imposed, drag-type force might act. Video pictures of the cell division process show that this in fact is true for cells dividing in static culture and even shortly after the onset of flow. However, for EC which have elongated in shape in response to flow, the cell divides into two elongated daughter cells while maintaining a relatively low, flattened profile throughout the entire process of cell division. (Ziegler et al., 1990).

Finally, the presence of flow has been shown to affect the expression of proto-oncogenes which regulate cell cycle. The stimulating effect of α-thrombin on c-myc expression was seven fold higher for EC in static culture as compared to EC preconditioned for 24 hours by steady flow (preliminary data)(Berk et al., 1991).

PDGF B-CHAIN EXPRESSION

To determine the effect of shear stress on growth factor production by EC, we studied the effect of shear stress (30 dyne/cm^2) on the expression of platelet derived growth factor (PDGF) A- and B-chain mRNA and, as a control, on glyceraldehyde-3-phosphate-dehydrogenase (GAPDH) mRNA (Mitsumata et al., 1991b). BAEC in a confluent monolayer (3 days after plating) were subjected to shear stress for 24 hours, after which total RNA was isolated and Northern blot analysis performed. After 24 hours of exposure to shear stress, PDGF B-chain expression was increased by a factor of ten over that observed for control static culture EC, while GAPDH expression was decreased. PDGF A-chain expression was not affected by flow. These data suggest that EC, under physiological shear stress, may differentially regulate gene expression. The decreased GAPDH expression would be consistent with attainment of a growth arrested state, while the increase in PDGF B-chain expression may be seen as a physiologic adaptation to flow.

Protein kinase C (PKC) was not involved in this flow response even though phorbol 12-myristate 13-acetate (PMA) induced PDGF B-chain mRNA expression. Activation of PKC alone, however, was insufficient to induce PDGF B-chain mRNA because the selective PKC activator, 1-ol.eoyl-2-acetyl-sn-glycerol, did not induce PDGF expression. A PKC-independent pathway was suggested by the fact that inhibition of PKC (downregulation with phorbol 12,13 dibutyrate or exposure to staurosporine) failed to block PMA or flow-induced PDGF B-chain expresion. These results demonstrate that flow-induced PDGF B-chain expression in EC appears to be mediated by a PKC-independent pathway.

SIGNAL RECOGNITION AND TRANSDUCTION

An important, unanswered question is how do EC recognize their flow environment, and having done so, how is this signal transduced into the changes in structure and function observed? There have been a number of studies of the mechanisms involved in this, and these initially have focused on the latter, i.e. the second messengers associated with the transduction of a flow signal. Such studies indicate that shear does stimulate the phosphoinositide system (Nollert et al., 1990; Prasad et al., 1989), that there is an elevation in intracellular calcium (Ando et al., 1988; Mo et al., 1991; Dull and Davies, 1991; Shen and Dewey, 1992; Geiger et al., 1992), and that there is a translocation of PKC from cytosol to membrane (Girard and Nerem, 1990).

This all suggests that the second messengers, known to be activated by hormonal agonists, also are stimulated by flow-induced stresses. However, how do EC recognize the flow environment in which they reside? There are many possibilities. These include mechanically-activated ion channels, the shear rate control of the transport of a small molecule through convection-diffusion coupling, an effective "strain gauge" which senses deformation of the cell's cytoskeletal structure, and a "shear" receptor in the cell's membrane, coupled to a G protein, but one which is shear sensitive.

The issues of signal recognition and transduction are not ones confined to just the influence of shear stress. The endothelial cell, as noted earlier, "sees" pressure and "rides" a basement membrane which is being cyclically stretched. Studies on the influence of cyclic stretch indicate important alterations in endothelial structure and function

DISCUSSION

There thus is clear evidence that the flow environment in which an endothelial monolayer resides participates in the regulation of the EC growth program. This is true of its intrinsic program, i.e. the regulation of the endothelial cell's own growth, as our results demonstrate that not only is the rate of EC proliferation decreased by flow, but entry into cell cycle is inhibited. If PDGF B-chain can be used as an example, there also is evidence of an influence of flow on the regulation of an EC's extrinsic growth program, i.e. the ability of an EC to influence the growth characteristics of neighboring cells. Furthermore, at least some insight has been gained into the signaling mechanisms involved in this response of EC to flow.

A limitation of these results is that, with one exception, they have all been obtained under conditions of steady flow, whereas *in vivo* the flow is pulsatile. The one exception is the experiment noted in which EC were exposed to a non-reversing, pulsatile flow; however, the observation that proliferation is further decreased compared to that for steady flow suggests that further experiments using a variety of different pulsatile flow environments (Helmlinger et al., 1991) might be illuminating.

Another limitation is that *in vivo* the endothelium is not only exposed to flow and the associated shear stress, but also to the cyclic stretch of the basement membrane on which the EC ride. As already noted, it has been demonstrated that there are alterations in EC biology due to the effect of cyclic stretch (Nerem and Girard, 1990).

Finally, a limitation is that the studies reported here have been conducted with endothelial monolayers grown on a solid substrate where, in contrast with the *in vivo* EC environment, there is no underlying smooth muscle cells (SMC). Yet the presence of neighboring cells is known to influence EC function(Stewart et al., 1990; D'Amore et al., 1987). Thus, to simulate the EC environment *in vitro* in a more physiologic manner will require a model in which an EC-SMC co-culture can be studied under conditions of pulsatile flow. This is the direction the work of our laboratory is now taking.

ACKNOWLEDGEMENT

This brief review represents research conducted with the support of National Science Foundation Grants ECS-8815656 and BCS-9111761.

REFERENCES

Ando J, Komatsuda T, Kamiya A (1988): Cytoplasmic calcium responses to fluid shear stress in cultured vascular endothelial cells. *In Vitro Cell Dev Biol* 24: 871-877.

Berk BC, Girard PR, Mitsumata M, et al. (1991): Shear stress alters the genetic program of cultured endothelial cells. In: *Proceedings of First World Congress of Biomechanics*, LaJolla, CA, August 30-September 4, p. 315.

Caplan BA and Schwartz CJ (1973): Increased endothelial cell turnover in areas of *in vivo* Evans blue uptake in pig aorta. *Arteriosclerosis* 17: 401-417.

D'Amore PA, Orlidge A and Herman IM (1987): Growth Control in the retinal microvasculature. In: *Progress in Retinal Research*, 7: 233-258, Osborne N and Chader G, eds. Pergamon Press, New York.

Dull RO, Davies PF (1991): Flow modulation of agonist (ATP)-response (Ca^{++}) coupling in vascular endothelial cells. *Am J Physiol* 261: H149-154.

Geiger RV, Berk BC, Alexander RW, Nerem RM (1992): Flow-induced calcium transients on single endothelial cells: spatial and temporal analysis. *Am J Physiol: Cell Physiol* 262:C1411-C1417.

Girard RB, Nerem RM (1990): Role of protein kinase C in transduction of shear stress to alterations in endothelial cell morphology. *J Cell Biochem* 14E: 210 (abstract).

Helmlinger G, Geiger RV, Schreck S, et al.(1991): Effects of pulsatile flow on cultured vascular endothelial cell morphology. *J Biomech Engr* 113: 123-131.

Jones PA (1982): Construction of an artificial blood vessel wall from cultured endothelial and smooth muscle cells. *J Cell Biol* 74: 1882-1886.

Kim DW, Gotlieb AI, Langille BL (1989): In vivo modulation of endothelial F-actin microfilaments by experimental alterations in shear stress. *Arteriosclerosis* 9: 439-445.

Levesque MJ, Liepsch D, Moravec S, and Nerem RM. (1986): Correlation of endothelial cell shape and wall shear stress in a stenosed dog aorta. *Arteriosclerosis* 6: 220-229.

Levesque MJ, Sprague EA and Nerem RM (1990): Vascular endothelial cell proliferation in culture and the influence of flow. *Biomaterials*, 11(9): 702-707.

Mitsumata M, Nerem RM, Alexander RW and Berk BC (1991a): Shear stress inhibits endothelial cell proliferation by growth arrest in the G_0/G_1 phase of the cell cycle. *FASEB J* 5(4): A527 (abstract).

Mitsumata M, Nerem RM, Alexander RW, and Berk, BC (1991b): Inverse relationship in mRNA expression between c-sis and GAPDH in endothelial cells subjected to shear stress. *Abstract book of workshop on Mechanical Stress Effects on Vascular Cells*, Atlanta, GA, April 20-21.

Mo M, Eskin SG, Schilling WP (1991): Flow-induced changes in Ca^{2+} signaling of vascular endothelial cells: effects of shear stress and ATP. *Am J Physiol* 260: H1698-H1707.

Nerem RM and Girard PR (1990): Hemodynamic influences on vascular endothelial biology. *Toxic Path* 18(4): 572-582.

Nollert MU, Eskin SG, McIntire LV (1990): Shear stress increases inositol trisphospate levels in human endothelial cells. *Biochem Biophys Res Commun* 170: 281-287.

Prasad ARS, Nerem RM, Schwartz CJ, et al. (1989): Stimulation of phosphoinositide hydrolysis in bovine aortic endothelial cells exposed to elevated shear stress. *J Cell Biol* 109: 331a (abstract).

Schwartz CJ, Sprague EA, Fowler SR, et al. (1983): Cellular participation in atherogenesis: Selected facets of endothelium, smooth muscle, and peripheral blood monocyte. In: *Fluid Dynamics as a Localizing Factor for Atherosclerosis*. G Schettler, RM Nerem, H Schmid-Schönbein, H Mörl and C Diehm (eds). Springer-Verlag, Heidelberg FRG, pp. 200-207.

Shen J, Dewey CF Jr. (1992): Fluid shear stress modulates cytosolic free calcium in vascular endothelial cells. *Am J Physiol: Cell Physiol* 262: C384-390.

Stewart DJ, Langleben D, Cernacek P and Cianflone K (1990): Endothelin release is inhibited by coculture of endothelial cells with cells of vascular media. *Am J Physiol* 259: H1928-H1932.

Weinberg CB and Bell E (1986): A blood vessel model constructed from collagen and cultured vascular cells. *Science* 231: 397-399.

Wright HP (1968): Endothelial mitosis around aortic branches in normal Guinea pigs. *Nature* 220: 78-79.

Ziegler T, Girard PR and Nerem RM (1990): Division of cultured endothelial cells in the presence of flow. AIChE 1990 Annual Meeting, Chicago, IL, Novmeber 11-16.

Zilla PP, Fasol RD and Deutsch M (eds) (1987): Endothelialization of vascular grafts. Karger, Basel, Switzerland.

DEFORMATION OF CHONDROCYTES WITHIN THE EXTRACELLULAR MATRIX OF ARTICULAR CARTILAGE

Van C. Mow
Orthopaedic Research Laboratory
Columbia University New York, New York 10032

Farshid Guilak
Musculo-Skeletal Research Laboratory
State University of New York at Stony Brook
Stony Brook, New York 11794

ABSTRACT

This study was aimed at determining possible mechanical signal transduction mechanisms in articular cartilage which may be responsible for the control of tissue remodeling. Based on existing data for articular cartilage deformational behavior, the extracellular and pericellular matrices, and the chondrocytes have all been modeled as biphasic materials, with distinct material properties. The cells are embedded and continuously bonded to the surrounding matrix. Finite element analysis of the stress, strain, fluid flow, hydraulic pressure and strain energy density were made on a configuration simulating an experiment where a cartilage explant was loaded in compression. All deformation fields were strongly dependent of chondrocyte material properties relative to the extracellular matrix. The existence of a pericellular matrix seems to reduce the stresses and strains the chondrocyte is subjected to. Confocal microscopy was used to determine the three-dimensional shape and organization of undeformed and deformed chondrocytes from mid-zone cartilage.

Microscopy data indicate that chondrocytes do not deform in an identical manner to the extracellular matrix, thus justifying the distinct material properties assigned to them in the finite element model. This finding led to our proposing an inverse method for determining chondrocyte material properties *in situ*. This method is based on shape-fitting the observed deformed chondrocyte with the finite element model predictions using an optimization algorithm. Further studies are required to assess the feasibility of this inverse method. Knowledge of the deformational field around the chondrocytes *in situ* is required for the understanding of mechanical signal transduction mechanisms responsible for mediating the cellular remodeling processes in this tissue.

INTRODUCTION

In solving engineering problems, we often perform analyses to determine the internal stresses and strains that exist within a structure to see if the structure can withstand the loads it is expected to support. This means that we must know its size and shape, the material it is made of, the external loads acting on it (surface tractions or body forces, e.g., gravity) and the initial conditions (whether it is stress free or whether there is a non-zero residual stress state). For stress-strain analyses of biological structures over short time intervals, many aspects of the traditional methods embodied in continuum mechanics (elasticity, viscous fluid mechanics, viscoelasticity, non-Newtonian fluid mechanics, plasticity, mixture theories, etc.) are perfectly adequate. Due to cellular remodeling processes within the tissue, Fung and co-workers have convincingly pointed to the uncertainties involved in assuming an initial zero stress state (Chuong, 1986; Fung, 1985; Fung, 1988). Based on the existence of a non-zero residual stress state, Fung (1985) proposed a new hypothesis regarding biological tissue function: "Each organ operates in a manner to achieve optimal performance in some sense. In particular the [non-zero] residual stress in the tissue distributes itself in a way to assure such performance. The optimal condition may vary from organ to organ; but, in general, it is not the same as zero residual stress when all the external loads are removed." Thus it is important to ask how does residual stress come about in a tissue? Fung argues that cells would remodel a region of the tissue around themselves if the stress concentration in the region is beyond some normal range of values by adding mass (collagen, elastin, proteoglycan) to bear load, thus reshaping and otherwise remodeling the tissue to change the state of stress. Conversely, if there is a general lowering of stress, the cells would resorb the unnecessary material. These hypertrophic and atrophic changes are well known to biologists and physicians (Stockwell, 1987; Woo, 1988).

Similar structure-function relationships and cellular remodeling processes also exist for bone. Meyer (1867) pointed out that the trabecular architecture of bone must have a mechanical basis, and that its pattern is similar to the principal stress trajectories in cantilever beams which he found in a 1866 statics book published by Culmann (1866). Wolff (1892) claimed to have proved the "law of bone transformation" and this is currently known as Wolff's law of bone remodeling. Roux (1895) introduced the concept of functional adaptation as a self-regulating mechanism of biological tissues which is controlled by mechanical stimulation. Since these early beginnings, many biomechanicians of modern times have attempted to discover the biologic mechanisms underlying these remodeling phenomenon (e.g., Cowin, 1986; Huiskes, 1991; Carter, 1987; Fung, 1990; Goldstein, 1987).

Other biologists and bioengineers have been interested in the effects that mechanical and electrochemical environments (stress, strain, hydrostatic and osmotic pressures, streaming potential, pH, ion concentration, etc.) have on modulating cellular activities and on signal transduction mechanisms within the tissue. For example, Folkman and Moscona (1978) have implicated cell shape, and in a parallel manner, Stockwell (1987) and Guilak and co-workers (1990) have suggested that cell deformation is important in controlling growth and metabolic events (Folkman, 1978; Guilak, 1990a; Guilak, 1990b). Additionally, some cell biologists have associated changes in cell shape with changes in metabolic activity and proliferation (e.g., Newman, 1988; Srivastava, 1974). Other investigators have suggested that hydrostatic and osmotic pressures (Hall, 1991; Schneiderman, 1986), and static and cyclic deformations (Guilak, 1992a; Guilak, 1992b; Sah, 1989; ,Terracio, 1988) can stimulate cellular responses. However, few studies have focused on the calculation of stresses and strains experienced by cells or in the immediate environment around the cell (Guilak, 1992a; Guilak, 1992b; Guilak, 1990a, Guilak, 1990b). Beside the notion of residual stress, there are a number of other very fundamental and difficult technical problems which must be overcome before some of these studies can be undertaken in a reliable and consistent manner. For example, little is known of the stress-strain relationship for cells and cell membranes, nor the manner with which cells are attached to its immediate environment; Do cells form a continuous bond with the ECM, or are the attachment points discrete? Do transmembrane and cell adhesion molecules (integrins) have a role in transmitting extracellular events into intracellular signals (Sundqvist, 1976)? Do tensegrity structures play roles in this signal transduction mechanism within the cell (Ingber, 1985; Ingber, 1992)?

Thus, in the recent tissue engineering literature, three fundamental questions have been raised regarding the theoretical formulation and experimental verification of problems of tissue growth and remodeling:

1) What are the physical mechanisms giving rise to residual stress in tissues? Related to this, are these mechanisms the same for all tissues and organs?

2) How can one incorporate these residual stresses as initial conditions, given the nonlinear nature of almost all biologic tissues, into an a stress-strain analysis of the tissue or organ simulating physiologic loading conditions? (This is necessary to verify Fung's hypothesis.)

3) How can one experimentally measure the cellular deformations *in situ*? Related to this question, what simplifying assumptions must be made to develop, and how can one determine, the constitutive laws for cells *in situ*?

For a specific tissue, articular cartilage, we have addressed the first question. Recently we have developed a swelling theory which provides a mechanism for the generation of the residual stress in cartilage (Lai, 1992), and we have a quantitative analysis of the residual stress and strain fields in the tissue (Setton, 1992). In this paper, we wish to report on a study addressing the third question, however, without a consideration of the residual stress. This will be left for future investigations.

ARTICULAR CARTILAGE AS A PARADIGM FOR STUDYING CELLULAR DEFORMATION AND TISSUE REMODELING

This study is specifically aimed at understanding the details of the articular cartilage remodeling process. Historically it has been observed that mechanical forces are capable of inducing remodeling changes in the structure of diarthrodial joints. Recently, studies have shown that changes in the mechanical environment of articular cartilage, such as those caused by joint disuse or overuse *in vivo*, will produce significant changes in its structure, composition, and mechanical properties (Tammi, 1987). Under normal physiological conditions, the metabolic activity of the chondrocytes maintains the tissue in homeostasis through a balance between the release of degradation products from the tissue and the synthesis of new components. To better understand these catabolic and anabolic events and how changes in the mechanical environment of the tissue may affect these biologic events, a number of *in vitro* studies have been performed on cartilage explants (e.g., Guilak, 1992b; Sah, 1989). In general, these studies have shown that static compression of cartilage explants can suppress proteoglycan and protein synthesis and release rates, while cyclic compression at specific frequencies and magnitudes can stimulate matrix synthesis rates. Though the signal transduction mechanisms of these responses are not yet understood, it is believed that the chondrocytes perceive stress, strain, osmotic and hydrostatic pressures, streaming potential, pH, and other signals, and subsequently respond by regulating their metabolic activity (Stockwell, 1987). In this study, in order to isolate the mechanism by which chondrocytes respond to this diverse array of signals, as a first step, focus is placed on how chondrocytes convert *mechanical* signals to a biochemical response. To do this, it is necessary to determine not only the deformation of the tissue as a whole, but also the details of the states of stress and strain around the cells. Thus it is necessary to determine: 1) the response of the ECM to static, cyclic, or intermittent loading or deformation histories such as those commonly used explant models; 2) the deformation history of chondrocytes during compression of the tissue; and 3) the nature of the mechanical interactions between the cell, the pericellular matrix (PCM), and the ECM. To completely formulate the problem, the mechanical properties and constitutive laws of the chondrocyte, the ECM, as well as those of the PCM, must be known.

General Procedure

The general procedure which we employ to determine chondrocyte material properties and deformation utilizes finite element modeling (FEM) and confocal scanning optical microscopy (CSOM). First, using finite element modeling, the spatial and temporal mechanical response of the ECM will be described under a variety of mechanical loading configurations (Guilak, 1990b). In other studies, a linear biphasic FEM has been used to describe the mechanical response of the tissue including structurally distinct inclusions such as chondrocytes (Guilak, 1990a). In this manner, the effects of various morphological characteristics, such as cell shape, intercellular spacing or cell location (from the surface), has been parametrically examined. Second, using CSOM, the three-dimensional changes in cell shape and volume will be measured during compression of the matrix (Guilak, 1992a). Finally, using finite element optimization methods, the FEM predictions of cell shape and experimental measurements from CSOM can be combined to determine the intrinsic material properties of individual chondrocytes.

FINITE ELEMENT MODELING OF CELL-MATRIX INTERACTIONS

Since no information on the intrinsic mechanical properties of chondrocytes *in situ* is available, it is assumed that they exhibit viscoelastic behaviors similar to other mammalian cells (Dong, 1988). In this study, it was assumed that, because of its thickness and structure, the cell membrane does not play a structural role in the mechanical behavior of the chondrocyte, but that the cytoskeleton provides the main elastic framework of the cell. It was also assumed that the fluid-solid interactions between the cytoskeleton and cytosol are responsible for the viscoelastic behavior of the cell. This notion is similar to the fundamental ideas of the biphasic theory used for articular cartilage. Thus, the biphasic theory was adopted to model the cell as a fluid-solid inclusion, embedded in and attached to the surrounding biphasic ECM. While the intrinsic properties (aggregate modulus, Poisson's ratio and permeability) defining the cell are assumed to be distinct from those of the ECM, it is assumed that, due to their small volume fraction in the tissue (1-10%), the cells do not contribute appreciably to the mechanical behavior of the tissue as a whole. Even for this simplistic model, the complexity of the governing equations of the biphasic theory precludes any analytical solution to problems. Accordingly, the biphasic finite element programs developed by Spilker et al. (1990) were used in the stress-strain analysis of this problem.

The major obstacle in the analysis of such a problem is the large difference in the geometric scales (two orders of magnitude) between the macro-scale problem (i.e., compression of the explant) and the micro-scale problem (i.e., mechanical interactions at the cellular level). By dividing the analysis into these two separate problems, a multiple scaling algorithm was developed to calculate the micro-scale response (Guilak, 1992a). In this algorithm, the FEM solution of the macro-scale problem was first used to determine the stresses, strains, fluid velocities, and pressures at each node in a model representing the entire explant (macro-scale). The results of this macro-scale FEM analysis were then used as the applied boundary conditions to a separate micro-scale FEM model. In this micro-scale FEM analysis, a much finer mesh was constructed representing the chondrocyte with its own distinct geometry and material properties. To match the micro-scale FEM problem to the macro-scale FEM problem, the time histories of the essential boundary conditions (fluid velocity and solid displacement) were calculated using the macro-scale model and these values were linearly interpolated across the faces of the micro-scale model. To use this technique, it is necessary to assume that the changes in stress and strain fields within the micro-scale mesh caused by the presence of the cell are not apparent at the edges of the mesh. This important assumption is valid at a distance of several cell diameters. However, the stress-strain field in the immediate vicinity of the cell is significantly affected, and these results define the stress-strain environment around the cell when the tissue is externally loaded.

This technique was used to examine the effects of several parameters, including relative material properties of the cell, PCM, and ECM, cell shape, and intercellular spacing of two axially aligned cells. Initially, the macro-scale problem was chosen to be that of an axisymmetric cylindrical disk of articular cartilage (5mm x 1mm) compressed under a step load using perfectly adhesive, permeable platens in a state of unconfined compression (Guilak, 1990b). A 469-node micro-scale model of the chondrocyte and its immediate surrounding ECM was then constructed, representing a 50 mm x 100 mm section at the center of the disk.

The material properties of the ECM were kept constant at a solid content of 0.17, aggregate modulus (H_A) of 0.7 MPa, Poisson's ratio (ν) of 0.125, and hydraulic permeability (k) of 7.6×10^{-15} $m^4/N \cdot s$. Cell elastic properties and permeability were varied by two orders of magnitude in each direction. Separate models were formed incorporating a PCM of thickness equal to 1/4 the diameter of the cell, and material properties of the PCM were similarly varied. The effects of cell shape were examined by varying the ellipticity (width/height) of the cell between 0.75 and 7.0. Interactions between cells were examined by using a model for two axially aligned spherical cells with various intercellular distances. The transient stress and strain fields of the cell were calculated by superimposing the stress due to deformation of the ECM, as calculated above, with the stress due to fluid pressure within the tissue. In this manner, the transient response in the vicinity of the cell, at any depth in the matrix, can be determined.

RESULTS OF FINITE ELEMENT ANALYSIS

Elastic Properties

Under axial compression of the ECM, the mechanical environment in the vicinity of the cell is highly nonuniform and dependent on the relative values of the elastic constants of the cell, the ECM, and the PCM. When the cell is less stiff than the surrounding ECM, a stress concentration occurs within the ECM at the cell-matrix junction. If the cell is modeled as being stiffer than the ECM, the highest principal stresses are within the cell. Changes in the relative Poisson's ratios of the cell (ν_c) and the ECM (ν_m) introduced significant changes in the radial stresses within the ECM. For example, if $\nu_c > \nu_m$, and is close to 0.5, the radial expansion of the cell under an axial compression of the ECM produces a large radial compressive stress in the ECM, while the cell itself experiences a large tensile strain in the radial direction. Conversely, a cell with ν_c approximately zero will experience both radial and axial compressive strain and a large decrease in its volume.

Cell Shape

FEM results showed larger principal stresses in the vicinity of the cell as the ellipticity of the cell was increased. Slightly prolate cells (ellipticity = 0.75) show even lower stress concentrations than spherical cells (Figure 1). Additionally, maximum principal strains within the cell increased significantly with increasing ellipticity. Correspondingly, the strain energy density within the cell increased significantly in oblate cells (ellipticity>1) as compared to more rounded ones (Figure 2).

Pericellular Matrix

The effects of the presence of a PCM were examined by parametrically assigning different material properties to a narrow region around the periphery of the cell. Interestingly, maximum principal stresses and strain energy density within the cell were decreased if the H_A of the PCM is either greater or less than that of the ECM (Figure 3). The PCM may also play a role in modulating the pH environment around the cell. Compaction of this region would change the fixed charge density as well as pH surrounding the cell. Both factors may be important in the signal transduction process required to stimulate cellular activities. Figure 4 shows the effects of (PCM stiffness)/(ECM stiffness) on axial stress and strain distributions along the radial axis. Clearly, the PCM has a profound affect on the mechanical environment of the chondrocyte.

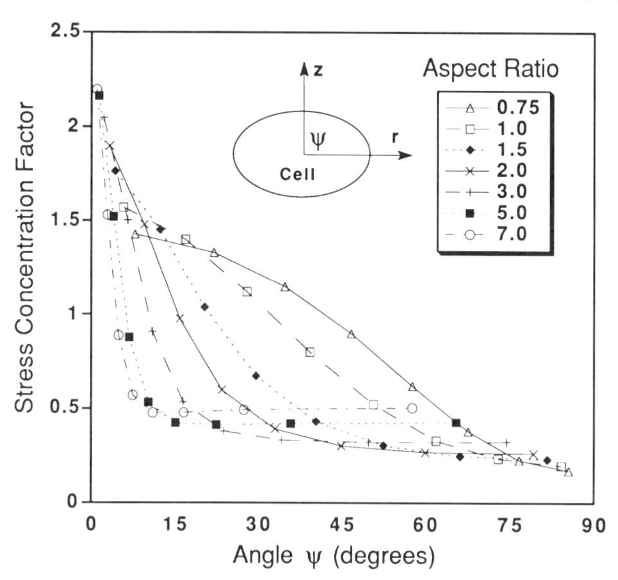

FIGURE 1. Axial stress concentration at the cell-matrix boundary for varying aspect ratio of the cell where $H_{A(CELL)} = (1/10)H_{A(ECM)}$. Both axial and radial stress concentration factors increase as the aspect ratio of the cell is increased.

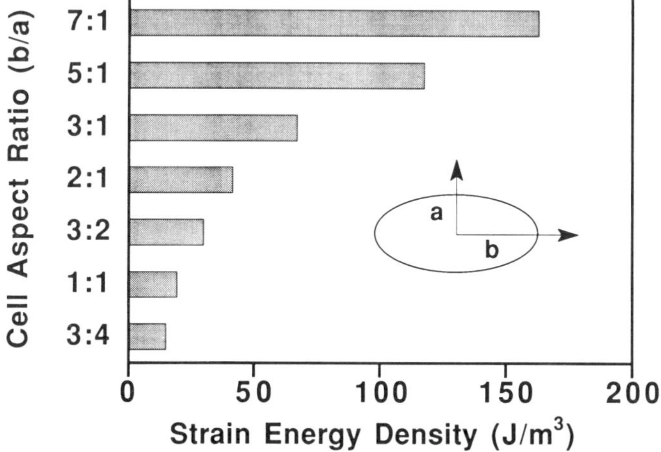

FIGURE 2. Strain energy density at the center of the cell increases significantly with increasing aspect ratio of the cell $H_{A(CELL)}=(1/10)H_{A(ECM)}$.

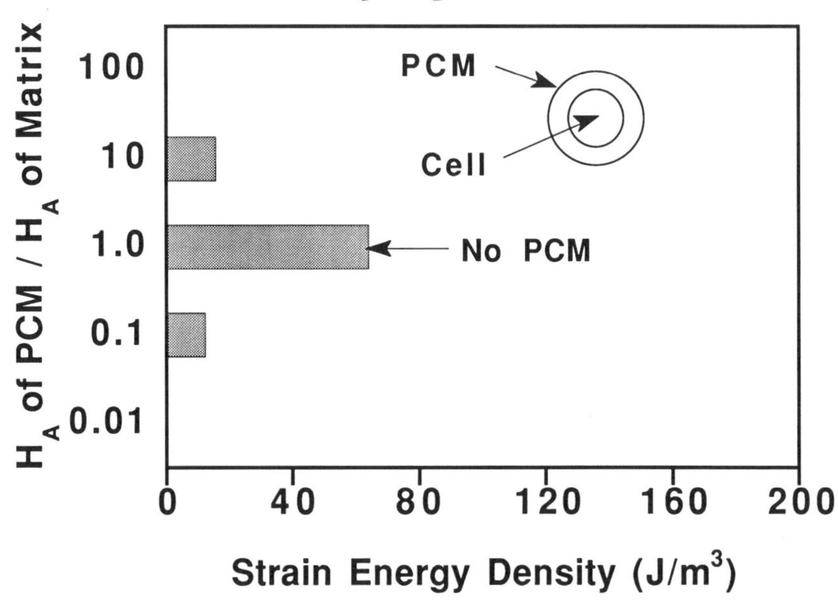

FIGURE 3. Strain energy density within the cells under axial loading decreases with any variation in the PCM properties ($H_{A(CELL)}=H_{A(ECM)}$).

Intercellular Spacing

Maximum principal stresses and strain energy density within the cell and the ECM vary insignificantly as intercellular spacing was varied from between 0.01 and 3.0 cell diameters. However, the overall stress-strain field forms a much more complex pattern due to cell-cell interactions. This may have an influence on the ultrastructural organization of the intercellular matrix.

Transient Behavior

Figure 5 shows the total axial stress on a spherical cell from the middle zone of the tissue cell following a sudden application of compressive loading on the tissue. The response of the ECM is characterized by high fluid exudation and strain at the top surface and pressurization of the fluid in the deep zones, followed by equilibration of these quantities through the depth (Mow, 1990). Consequently, stress levels in the solid ECM during the early time period after

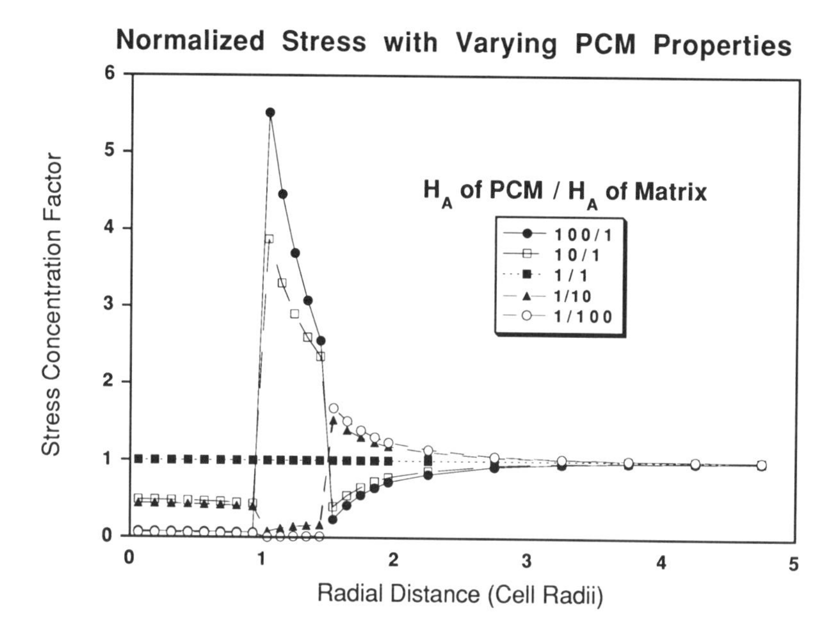

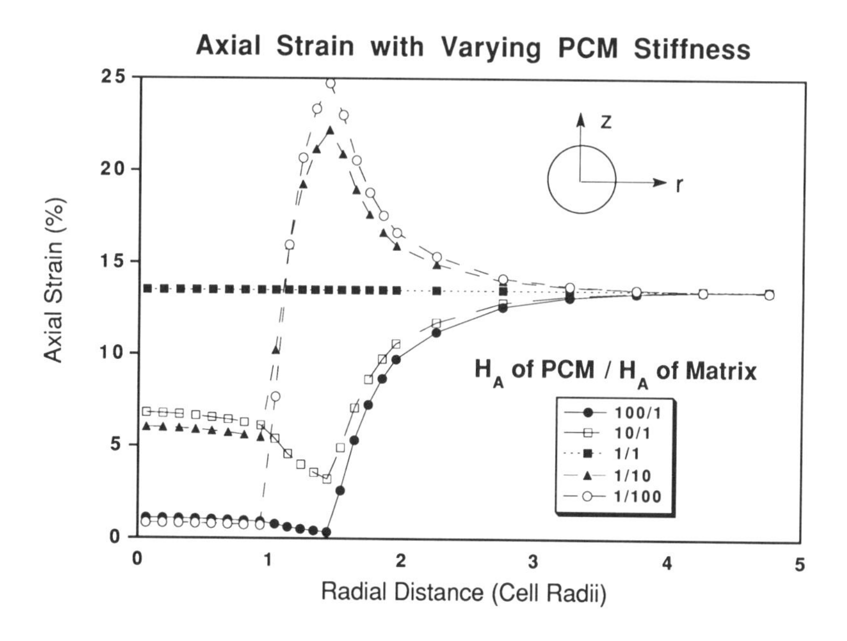

FIGURE 4. Axial stress (top) and strain (bottom) in the vicinity of the cell with varying pericellular stiffness. The axial stress and strain acting on the cell are decreased if the PCM stiffness is greater or less than the matrix stiffness ($H_{A(CELL)}=H_{A(ECM)}$).

loading are lower since most of the load is being carried by the hydraulic pressure in the fluid phase of the tissue. In time, the load is transferred to the solid ECM giving rise to higher stress levels (e.g., t=300s in Figure 5). Streamlines of interstitial flow under the same conditions are shown in Figures 6a & b. Parametric results indicate that a high cell permeability tends to channel interstitial flow through the cell while a low cell permeability tends to channel flow around the cell (Figure 6a). Figure 6b shows the evolution of interstitial fluid flow for t=2s, 6s, 16s and 30s; note v^{sf} denotes fluid flow relative to the porous-permeable solid matrix. The permeability of the PCM also affects interstitial flow in a similar manner. These results provide insight on how nutrient and ion transport around the cell may be affected during matrix compression.

EXPERIMENTAL METHODS: CONFOCAL MICROSCOPY

For the second portion of the study, CSOM was used to determine the deformation of live chondrocytes within an explant during compression. CSOM is a relatively new technique whereby numerous discrete optical sections can be taken through the depth of the tissue sample. These discrete sections can be merged to allow the formation of three-dimensional images of live cells *in situ.*

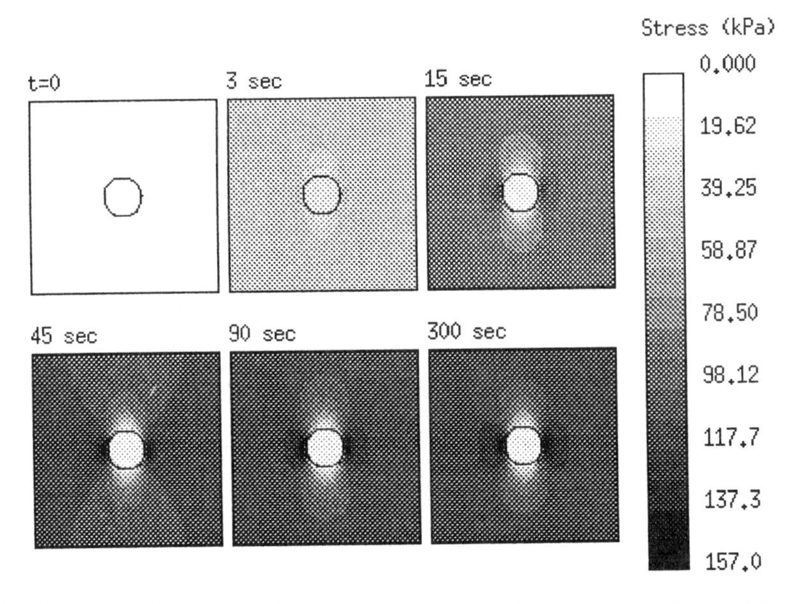

FIGURE 5. Maximum principal stresses (compressive) on a spherical cell in the middle zone of an explant under creep loading where H_A of the cell = 0.1 H_A of the ECM.

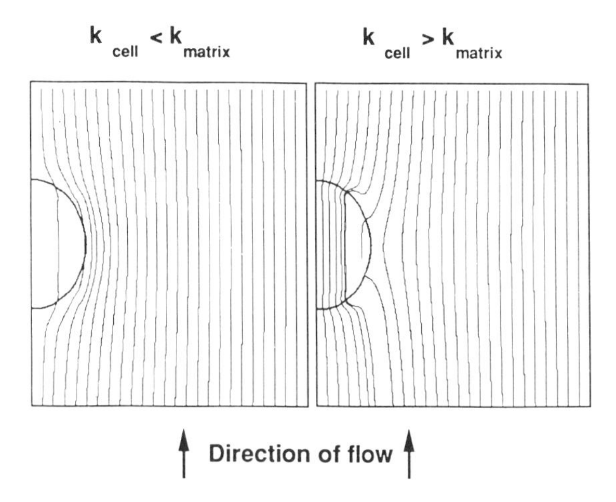

FIGURE 6a. Streamlines at t=2s showing the magnitude and direction of the interstitial velocity for the case where $k_{(CELL)} = 1/10k_{(ECM)}$ and $k_{(CELL)} = 10*k_{(ECM)}$, respectively. Interstitial velocity away from the cell is $1.25x10^{-6}$m/s.

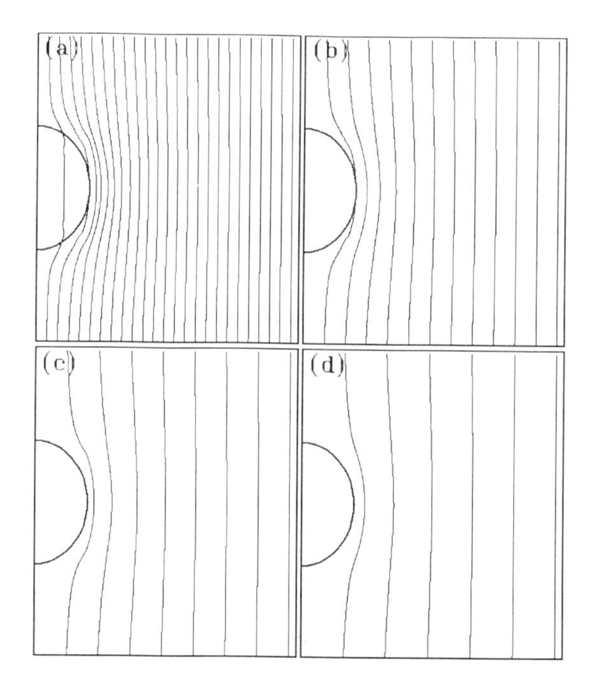

FIGURE 6b. Streamlines showing the magnitude and direction of the interstitial velocity for $k_{(CELL)} = 1/10k_{(ECM)}$ over time. (a) t=2s, uniform interstitial velocity (v^{sf})=$1.25x10^{-6}$m/s. (b) t=6s, v^{sf}=$6.5x10^{-7}$m/s. (c) t=16s, v^{sf}=$3.0x10^{-7}$m/s. (d) t=30s, v^{sf}=$1.5x10^{-7}$m/s.

In the experiment, cylindrical explants of articular cartilage were taken from the femoral condyles of freshly slaughtered cows and maintained in tissue culture until testing. For epifluorescent imaging of chondrocytes, tissue explants were incubated for 30 minutes in 1 μM solution of BCECF-AM, or in a 5μM solution of Cl-BODIPY, which are fluorescent cytosolic indicators (Molecular Probes Inc., Eugene OR). Samples were placed in a specially designed computer-controlled loading apparatus in the workspace of the microscope. The 3-D images of cellular patterns through the depth of the tissue were formed by recording 30 optical sections at an increment of 2 μm/section. Images were recorded at a resolution of 128 x 128 pixels using a 60x, 1.4NA objective. Volume images were recorded before compression and at 1200s following a 20% surface-to-surface compression.

Image data analysis and volume rendering were performed using the PV-WAVE CL software package (Precision Visuals Inc., Boulder CO). Morphometric measurements of individual cells were formed by decomposing the image of the chondrocyte surface into an array of approximately 2000 linked polygons. Vertices for these polygons were determined by performing an optical iso-intensity contour of the cell in three orthogonal directions. Geometric modeling algorithms were then used to produce shaded surface solid models of the chondrocytes, as well as to calculate changes in cell volume and cell height with compression.

RESULTS OF CONFOCAL MICROSCOPY STUDIES

Compression of the live explant resulted in significant changes in the shape and volume of the chondrocyte *in situ* (see Table 1 and Figure 7). Figure 7 is a CSOM determined 3-D image of a mid-zone cartilage chondrocyte before and after compression. Table 1 shows the changes in dimensions and volume of six randomly selected cells. These findings are in general agreement with those of Freeman et al. (1991), who measured changes in cross-sectional area of chondrocytes embedded in an agarose matrix under compression. These deformational results imply a number of possible factors which may be involved in cellular mechanical signal transduction. Changes in cell shape may result in stretching of the cell membrane, where stretch-activated ion channels have been identified (Guharay, 1984). Alternatively, deformation of the membrane and cytoskeletal structures may be involved in an intracellular signalling process (Ingner, 1985; Ingber, 1992; Sundqvist, 1976; Terracio, 1988).

Change in cell volume (ΔV_c) which differs from the change in the overall explant volume (ΔV_e) implies that fluid flow is occurring across the cell membrane as the the tissue is compressed. Whether fluid is lost or gained depends on the actual compressive stiffness and Poisson's ratio of the cell relative to the ECM. While such plasma membranes are generally described as being highly permeable to water (Stryer, 1981), no measurements currently exist in the literature on their hydraulic permeability and thus the rate-limiting step in fluid flux. However, any such changes occurring in the fluid content of the cells

Cell #	Original Volume (μm^3)	Final Volume (μm^3)	Volume Change (%)	Original Height (μm)	Final Height (μm)	Height Change (%)
52.1	2136	1926	9.8	28	24	14
52.2	2328	2047	12.0	28	24	14
52.3	1976	1710	13.5	28	24	14
53.1	3777	3316	12.2	30	26	13
53.2	4531	4087	9.8	28	24	14
53.3	4218	3773	10.5	24	20	17
mean± s.d.	3161 ±1142	2809 ±1038*	11.3 ±1.5	27.7 ±2.0	23.7 ±2.0*	14.5 ±1.1

*$p<0.001$ vs. original volume/height, Student T-test

TABLE 1. Changes in cell height and volume with 20% compression of the matrix

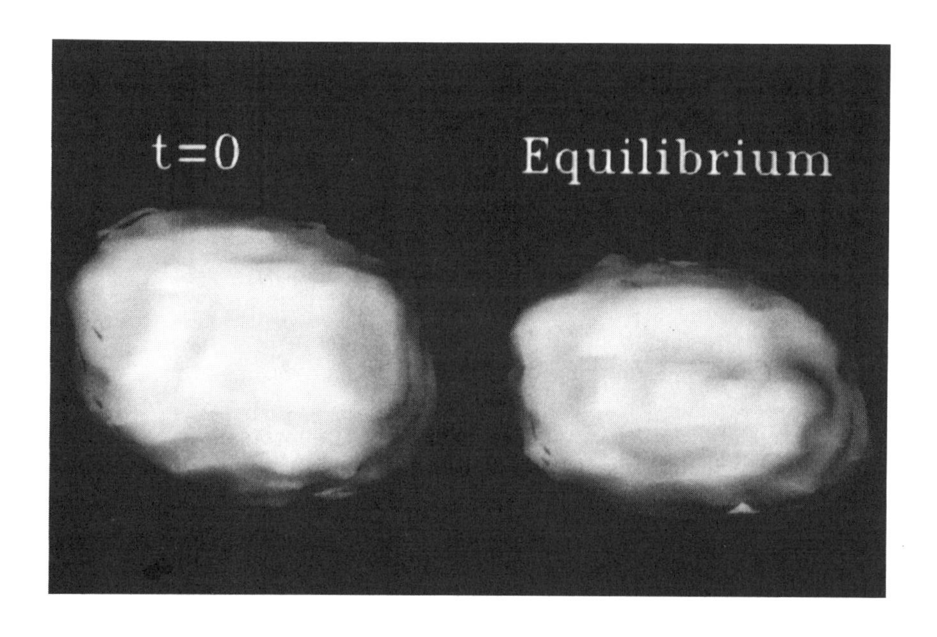

FIGURE 7. Three-dimensional shaded surface images of a chondrocyte from the middle zone of articular cartilage before (left) and after (right) 20% compression of the matrix.

and its surrounding environment may be responsible for variations in the ionic concentration in and around the cell, and could significantly affect the osmotic environment of the tissue. Electrokinetic effects may also be present due to streaming potentials induced by the fluid flow through the cell. Such phenomena may play a role in cellular signal transduction, and may be analyzed in more detail using the triphasic theory for charged hydrated soft tissues (Lai, 1992). Finally, the random selection of cells compression data in Table 1 shows that ΔV_c's are all less than ΔV_e; this implies that the cells are less compressible than the surrounding ECM. More data needs to be taken from other zones of the tissue, and from other tissues to verify if this result is true in general, or true specifically for the mid-zone of articular cartilage.

FINITE ELEMENT OPTIMIZATION

We have demonstrated that using the 3-D morphologic information from the CSOM, chondrocyte deformation may be measured *in situ* as a live explant is compressed. From CSOM, the undeformed and deformed shapes of the chondrocytes are known at a resolution of 0.5μm. Assuming the biphasic constitutive theory is applicable for the chondrocyte as well as for the explant, and assuming the macro-scale and micro-scale modeling procedure is valid, then by using an inverse method to match the FEM predicted deformed shapes of chondrocytes (parametric) and the CSOM observed shapes of deformed chondrocytes, it may be possible to calculate the mechanical properties (H_A, ν_s, k) of these cells *in situ*. This type of analysis can also be carried out using any of the different models proposed in this study to represent various chondrocyte morphologies (presence/absence of PCM, varying ellipticity of the cell, varying intercellular distance). To date, however, an optimization algorithm has not been developed, using the concept of an iterative scheme on shape-fitting, to determine the intrinsic mechanical properties of the chondrocytes.

DISCUSSION

In various studies, both the shape of the chondrocyte as well as its mechanical environment have been implicated as major factors in phenotypic expression. It is thus essential to understand changes in the mechanical environment of the chondrocyte and chondrocyte deformation itself before the signal transduction mechanism(s) of the chondrocyte can be isolated and fully understood. The approach presented in this study provides a combined theoretical and experimental method to study the mechanics of cell-matrix and cell-cell interactions in such tissues as articular cartilage. Results from our FEM analysis indicate that the mechanical environment around the cell is highly non-uniform and dependent on a number of factors such as cell shape, intercellular spacing, and cell material properties relative to the PCM and ECM. The presence of a PCM also significantly alters the mechanical response of the cell, and the results of our study imply a functional role for the PCM. CSOM studies indicate that chondrocytes undergo significant changes in their shape and volume during compression of the tissue. Together, these techniques provide for the first time a method of determining the intrinsic mechanical properties of chondrocytes *in situ*, and thus the complete mechanical environment of the chondrocyte. With further knowledge of the modes of signal conduction to the chondrocyte, tremendous potential exists for possible treatments of pathological conditions such as osteoarthritis which begins from articular cartilage degeneration.

CONCLUSION

We have determined that chondrocytes must be subjected to very complex stress and strain field in its immediate environment. The nature of the deformation field is unknown at present since no definitive determination of chondrocyte nor PCM properties have ever been made. These remain as worthwhile and challenging tasks for future investigation. An inverse method has been proposed using FEM of cell-matrix interactions, with presumed biphasic constitutive laws for the ECM and the chondrocyte, to determine cellular mechanical properties. Confocal microscopy has been used to define 3-D chondrocyte shapes and deformation in live explants. From these studies, it has been shown that mid-zone articular cartilage chondrocytes do not have the same deformational properties as the ECM surrounding it.

ACKNOWLEDGEMENT

This work was sponsored by a Frank E. Stinchfield fellowship (FG) from the New York Orthopaedic Hospital Research Laboratory at Columbia University.

REFERENCES

Carter DR, Fyhrie DP and Whalen RT (1987): Trabecular bone density and loading history: Regulation of connective tissue biology by mechanical energy. *J Biomechanics* 20:785-794.

Chuong CJ and Fung YC (1986): Residual stress in arteries. In: *Frontiers in Biomechanics,* Schmid-Schonbein GW, Woo SL-Y and Zweifach BW, eds. New York: Springer-Verlag.

Cowin SC (1986): Wolff's law of trabecular architecture at remodelling equilibrium. *J Biomech Engng* 108:83-88.

Culmann C (1866): *Die graphische Statik.* Meyer und Zeller, Zurich.

Dong C, Skalak R, Sung KLP, Schmid-Schonbein GW and Chien S (1988): Passive deformation analysis of human leukocytes. *J Biomech Engng* 110:27-36.

Folkman J and Moscona A (1978): Role of cell shape in growth control. *Nature* 273:345-349.

Freeman PM, Natarjan RN, Kimura JH and Andriacchi TP (1991): Chondrocytes respond to mechanical loading by maintaining a constant aspect ratio. *Trans Orthop Res Soc* 16:54.
Fung YC (1985): What principle governs the stress distribution in living organs. In: *Biomechanics in China, Japan and USA,* Fung YC, Fukada E and Wang JJ, eds. Beijing, China: Science Press.
Fung YC (1988): Cellular growth in soft tissues affected by the stress level in service. In: *Tissue Engineering,* Skalak R and Fox CF, eds. New York: Alan Liss Press.
Fung YC (1990): *Biomechanics: Motion, Flow, Stress and Growth.* New York: Springer-Verlag.
Goldstein SA (1987): The mechanical properties of trabecular bone: dependence on anatomic location and function. *J Biomechanics* 20:1055-1061.
Guharay F and Sachs F (1984): Stretch-activated single ion channel currents in tissue-cultured embryonic chick skeletal muscle. *J Physiol* 352:685-701.
Guilak F (1992a): Cell-matrix interactions and metabolic changes in articular cartilage explants under compression. Ph.D. Dissertation, New York: Columbia University.
Guilak F, Meyer BC, Ratcliffe A and Mow VC (1992b): Quantification of the effects of matrix compression on proteoglycan metabolism in articular cartilage explants. *J Orthop Res,* in review.
Guilak F, Ratcliffe A and Mow VC (1990a): The stress-strain environment around a chondrocyte: A finite element analysis of cell-matrix interactions. *Adv in Bioengng,* ASME 17:395.
Guilak F, Spilker RL and Mow VC (1990b): A finite element model of cartilage extracellular matrix response to static and cyclic compressive loading. *Adv in Bioengng,* ASME 17:225.
Hall AC, Urban JPG and Gehl KA (1991): The effects of hydrostatic pressure on matrix synthesis in articular cartilage. *J Orthop Res* 9:1-10.
Huiskes R (1991): Biomechanics of artificial joint fixation. In: *Basic Orthopaedic Biomechanics,* Mow VC and Hayes WC, eds. New York: Raven Press.
Ingber DE (1992): Mechanochemical switching between growth and differentiation by extracellular matrix: Possible uses of a cellular tensegrity mechanism. *J Cell Biochem,* (Supp.) 16F, 120.
Ingber DE and Jamieson JD (1985): Cells as tensegrity structures: architectual regulation of histodifferentiation by physical forces tranduced over basement membrane. In: *Gene Expression during Normal and Malignant Differentiation,* Anderson LC, Gahmberg CG and Ekblom P, eds. New York: Academic Press.
Lai WM, Hou JS and Mow VC (1992): A triphasic theory for the swelling and deformational behaviors of articular cartilage. *J Biomech Engng* 113:187-197.
Meyer GH (1867): Die architektur der spongiosa. *Archiv fur Anatomie, Physiologie, und wissenschaftliche Medizin,* (Reichert und wissenschafliche Medizin, Reichert und Du Bois-Reymonds Archiv) 34:615-625.

Mow VC, Hou JS, Owens JM and Ratcliffe A (1990): Biphasic and quasi-linear viscoelastic theories for hydrated soft tissues. In: *Biomechanics of Diarthrodial Joints,* Mow VC, Ratcliffe A and Woo SL-Y, eds. New York: Springer-Verlag.

Newman P and Watt FM (1988): Influence of cytochalasin D-induced changes in cell shape on proteoglycan synthesis by cultured articular chondrocytes. *Exp Cell Res* 178:199-210.

Roux W (1895): *Gesammelte Abhandlungen uber die entwicklungs mechanik der Organismen.* Leipzig: W Engelmann.

Sah RLY, Kim YJ, Doong JYH, Grodzinsky AJ, Plaas AHK and Sandy JD (1989): Biosynthetic response of cartilage explants to dynamic compression. *J Orthop Res* 7:619-636.

Schneiderman R, Keret D and Maroudas A (1986): Effects of mechanical and osmotic pressure on the fate of glycosamionglycan synthesis in human adult femoral head cartilage. *J Orthop Res* 4:393-406.

Setton LA, Gu WY, Lai WM and Mow VC (1992): Pre-stress in articular cartilage due to internal swelling pressures. *Adv in Bioengng,* ASME, In Press.

Spilker RL, Suh JK, Vermilyea ME and Maxian TA (1990): Alternate hybrid, mixed, and penalty finite element formulations for the biphasic model of soft hydrated tissues. In: *Biomechanics of Diarthrodial Joints,* Mow VC, Ratcliffe A and Woo SL-Y, eds. New York: Springer Verlag.

Srivastava VM, Malemud CJ and Sokoloff L (1974): Chondroid expression by rabbit articular cells in spinner culture following monolayer culture. *Conn Tiss Res* 2:127-136.

Stockwell RA (1987): Structure and function of the chondrocyte under mechanical stress. In: *Joint Loading: Biology and Health of Articular Structures,* Helmenin HJ, et al., eds. Bristol, U.K.: Wright-Butterworth Scientific.

Stryer L (1981): *Biochemistry,* 2nd Ed., San Francisco: WH Freeman.

Sundqvist KG and Ehrnst A (1976): Cytoskeletal control of surface membrane mobility. *Nature* 264:226-231.

Tammi M, Paukkonen M, Kiviranta I, Jurvelin J, Saamanen AM and Helmenin HJ (1987): Joint induced alteration in articular cartilage. In: *Joint Loading: Biology and Health of Articular Structures,* Helmenin HJ, et al., eds. Bristol, U.K.: Wright-Butterworth Scientific.

Terracio L, Peters W, Durig B, Miller B, Borg K and Borg TK (1988): Cellular hypertrophy can be induced by cyclical mechanical stretch *in vitro.* In: *Tissue Engineering,* Skalak R and Fox CF, eds. New York: Alan Liss Press.

Wolff J (1892): *Das Gesetz der Transformation der Knochen.* Berlin: Hirschwald.

Woo SLY and Buckwalter JA (1988): *Injuries and Repair of the Musculoskeletal Soft Tissue.* Park Ridge, Illinois: Amer Acad Orthop Surg.

MECHANICAL STRETCH RAPIDLY ACTIVATES MULTIPLE SIGNALING PATHWAYS IN CARDIAC MYOCYTES

Seigo Izumo and Jun-ichi Sadoshima
Molecular Medicine and Cardiovascular Divisions, Beth Israel Hospital, and Department of Medicine and Program in Cell and Developmental Biology, Harvard Medical School

INTRODUCTION

In living animals, many types of cells are normally exposed to a variety of internal and external forces. Although it is well known that mechanical forces cause a variety of effects on the structure and function of the cells, little is known as to how mechanical stimuli are converted into intracellular signals of gene regulation (reviewed in Ingber, 1991; Vandenburgh, 1992).

External load plays a critical role in determining muscle mass and its phenotype in both cardiac and skeletal muscles *in vivo* (for review see Morgan and Baker, 1991; Vandenburgh, 1992). The first direct evidence that muscle cells are able to sense the external load, in the absence of neuronal or hormonal factors, came from a study by Vandenburgh and Kaufman (1979) who demonstrated that cultured chick skeletal muscle cells grown on an elastic substrate underwent hypertrophy (increase in cell size without cell division) in response to static stretch of the substrate. Recently, we and others have shown that applying stretch to neonatal cardiac myocytes in culture results in a transcriptional activation of *c-fos* proto-oncogene and many other immediate-early genes and this was followed by an appearance of the hypertrophic phenotype (Komuro et al., 1990, 1991; Sadoshima et al., 1992a).

Several studies have reported changes in intracellular second messengers in response to a variety of mechanical stimuli (Vandenburgh, 1992). However, these studies were performed in diverse cell types using different means to apply mechanical stimuli and the physiological consequence of activation of these second messenger systems has not been fully elucidated. Therefore, it is not known whether a defined mechanical stimulus causes activation of a single or multiple second messenger systems in a given cell type and how these second messengers relate to alterations in gene expression in response to mechanical stimuli. To address this question, we performed a systematic analysis to identify the signal transduction pathways of the "mechano-transcription coupling" process using an *in vitro* model of stretch-induced cardiac hypertrophy (Sadoshima et al., 1992a). Because this model allows us to apply a simple and controlled mechanical stimulus to cardiac myocytes cultured in a defined serum free medium, it is a suitable system to dissect the signal transduction pathways of mechano-transcription coupling (Sadoshima et al., 1992b). We used stretch-induced *c-fos* gene expression as a "nuclear-marker" and examined the mechano-transcription coupling in a reverse way (Sadoshima and Izumo, 1993).

RESULTS

We first examined *cis*-acting elements which confer stretch-responsiveness to the *c-fos* gene promoter. The *c-fos* gene promoter is known to contain two major inducible *cis*-acting elements (Fig. 1A), the serum response element (SRE) and calcium/cAMP response element (Ca^{2+}/CRE) (Sheng and Greenberg, 1990). To identify the precise location of the stretch responsive element, six *c-fos*-CAT reporter constructs were transiently transfected into cardiac myocytes and induction of CAT activity by 2 h stretch was examined. As shown in Figure 1B, the -356wt construct containing the intact SRE exhibited significant stretch-inducible CAT activity, and the point mutations to the SRE (pm12 and pm18) abolished the stretch responsiveness in the *c-fos* promoter. This suggests that the stretch responsive element maps to the SRE. The mutated SRE in pm18 is known to bind serum response factor (SRF) but to fail to form the ternary complex with $p62^{TCF}$, resulting in the loss of response to protein kinase C (PKC) (Graham and Gilman, 1991; Sadoshima and Izumo, 1993). The failure of induction of CAT activity in pm18 suggests that the PKC- and $p62^{TCF}$-dependent pathway is essential in stretch-induced *c-fos* expression. The failure of induction in -71wt, pm18 and pm12 constructs which contain the intact Ca^{2+}/CRE suggests that the Ca^{2+}/CRE alone is not sufficient to confer stretch-responsiveness to the *c-fos* promoter. Another construct (SRE-56) in which one copy of SRE was ligated to a minimum *c-fos* promoter containing -56 bp upstream sequence also responded to stretch to a similar degree as -356wt construct (Fig. 1B), indicating that SRE is sufficient to confer the stretch responsiveness.

Measurement of PKC activity indicated that stretch caused a rapid doubling in PKC activity and this activation persisted for at least 30 min (Sadoshima and Izumo, 1993). Treatment of cells with protein kinase inhibitors H7 and staurosporine completely prevented stretch-induced *c-fos* expression. Downregulation of PKC activity by a 24 h treatment with phorbol 12 myristate 13 acetate (PMA) also blocked stretch-induced *c-fos* expression (data not shown). These results suggest that stretch activates PKC and PKC is necessary for stretch-induced expression of *c-fos* genes, consistent with the DNA transfection experiments.

We next examined how PKC is activated by cell stretch. It has been shown that diacylglycerol (DAG) is an endogenous activator of PKC (Nishizuka, 1992). One of the major pathways of DAG formation is hydrolysis of phosphatidylinositol bisphosphate (PIP_2) by activation of phospholipase C (PLC), which results in production of DAG and inositol trisphosphate (IP_3) (Berridge et al., 1984). Measurement of IP_3 indicated that stretch caused a 3.5-fold increase in IP_3 within 1 min (Fig. 2A). D 609, a specific inhibitor of PLC (Schalasta and Doppler, 1990), prevented IP_3 accumulation by stretch (Fig. 2A) and significantly suppressed stretch-induced *c-fos* expression (data not shown). These results are consistent with the notion that PLC activation is likely to be an important step in mediating the stretch response.

A

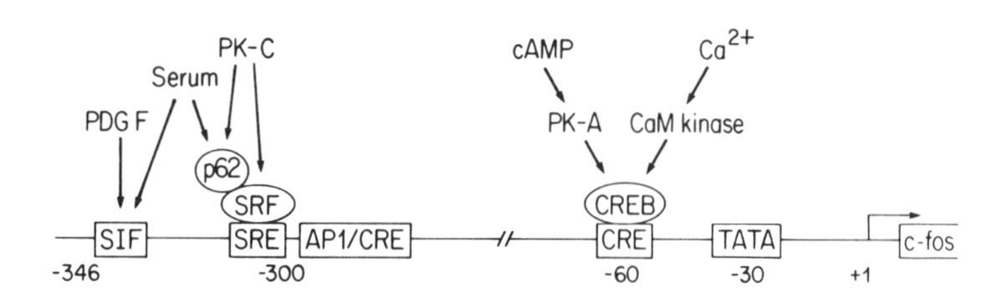

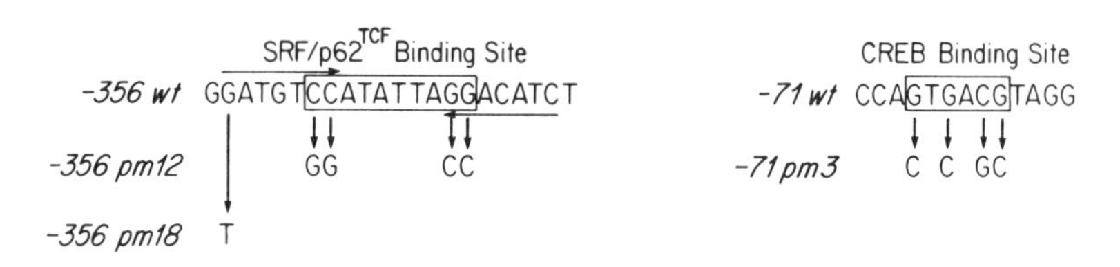

B

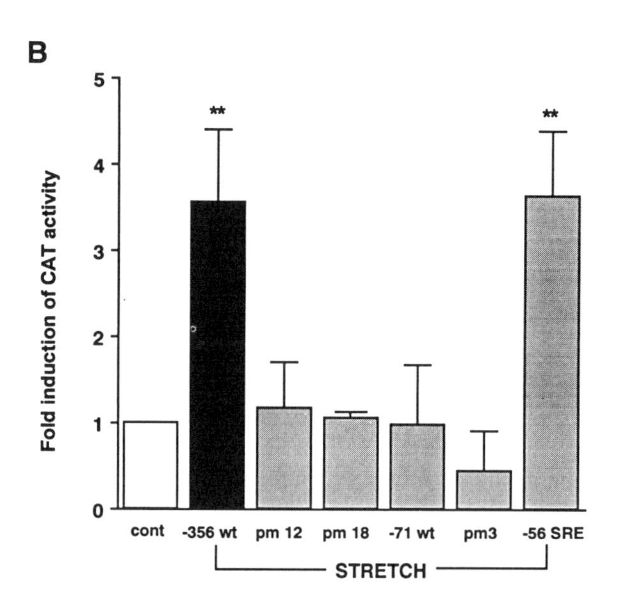

FIGURE 1. Relative stretch-induced CAT activity in various *c-fos* CAT constructs. **(A)** Schematic representation of the mouse *c-fos* promoter region. **(B)** Forty eight h after transfection, a two h stretch was applied. For each construct, stretch-induced CAT activity was normalized to that of control without stretch (cont). Data are mean $\pm$ SE of 3 to 5 independent transfection experiments. CAT, chloramphenicol acetyltransferase; PK-C, protein kinase C; PK-A, protein kinase A; CaM kinase, Ca^{2+} calmodulin dependent protein kinase; PDGF, platelet derived growth factor; SIF, *sis*/PDGF-inducible element; SRF, serum response factors; CREB, CRE binding proteins; cont, control; **$p<0.01$ vs non-stretched control.

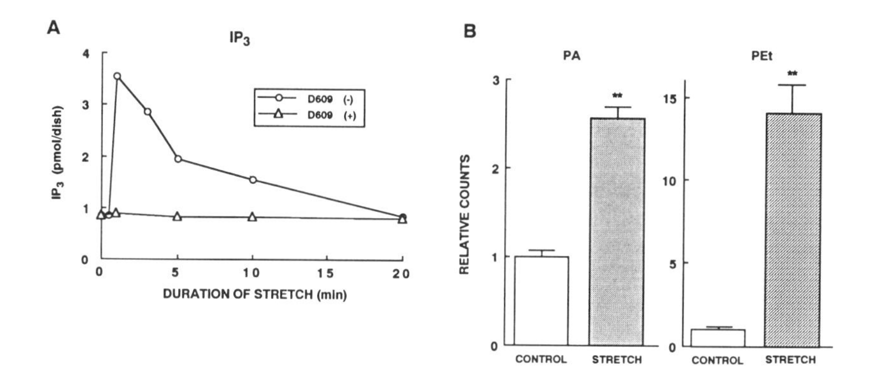

FIGURE 2. Cell stretch activates PLC and PLD. **(A)** Stretch-induced production of IP_3. Cardiac myocytes were stretched for the indicated times in the presence (triangles) of or absence (circles) of D609 (100µM), an inhibitor of PLC. **(B)** Stretch-induced hydrolysis of phosphatidylcholine. Phosphatidic acid (PA) and phosphatidyletanol (PEt) were separated by TLC.

It is known that another major membrane phospholipid, phosphatidylcholine is also hydrolyzed by phospholipases C, D and A_2 in response to various growth stimuli, and their breakdown products also act as second messengers (Nishizuka, 1992). Phosphatidic acid (PA), produced through activation of phospholipase D (PLD), is metabolized into DAG by PA phosphohydrolase, and can be the major pathway for the activation of PKC in some systems (Exton, 1990). In cardiac myocytes prelabeled with $[^3H]$ myristic acid, stretch in the presence of 0.5 % ethanol resulted in a 2.5-fold increase in PA and a 14-fold increase in radiolabeled phosphatidylethanol (PEt), a specific marker of PLD activation (Fig. 2B). The results indicate that stretch significantly activates PLD. Consistent with activation of both PLC and PLD, measurement of DAG indicated that stretch caused increase in cellular DAG content (150%) which persisted for more than 30 min (data not shown).

Arachidonic acid and its metabolites act as second messengers and regulate activity of ion channels and protein kinases in many systems (reviewed in Piomelli and Greengard, 1990). We found that 3 min of stretch caused 4-fold increase in release of $[^3H]$ arachidonic acid and its metabolites from cardiac myocytes prelabeled with $[^3H]$ arachidonic acid. The release of $[^3H]$ arachidonic acid and its metabolites was inhibited by pretreatment with quinacrine, a putative PLA_2 inhibitor. Stretch-induced *c-fos* expression was also attenuated by 60% with quinacrine (data not shown).

It has been shown that many cell growth stimuli cause tyrosine-phosphorylation of various intracellular substrates, through receptors that are either directly or indirectly coupled to tyrosine kinases (Cantley et al., 1991). Anti-phosphotyrosine immunoblotting showed stretch causes a marked increase in phosphotyrosine content of several cellular proteins, including 42-44 kDa proteins (data not shown). Immune complex kinase assay indicated that stretch activated p42 and p44 mitogen activated protein (MAP) kinase (Fig. 3A). It is known that MAP kinases phosphorylate one of the S6 kinases, RSK ($pp90^{RSK}$), and regulate its activity (Blenis, 1991). An immune-complex S6 peptide kinase assay indicated that stretch activated $pp90^{RSK}$ (Fig. 3B).

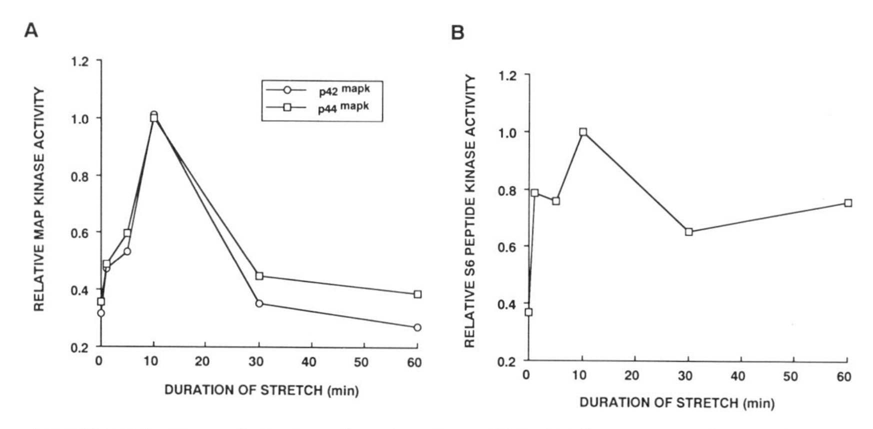

FIGURE 3. Stretch-induced activation of MAP kinases and S6 peptide kinases. **(A)** Time course of kinase activity of p42 and p44 MAP kinases measured by immune complex in-gel kinase assay using myelin basic protein as substrate. **(B)** Time course of stretch-induced activation of S6 peptide kinase activities.

Small GTP-binding proteins encoded by the *ras* proto-oncogenes ($p21^{ras}$) have been shown to play an important role in tyrosine kinase signal transduction pathways (Satoh *et al.*, 1992). Determination of guanine nucleotide bound to ras indicated that stretch caused a 2.5-fold increase in GTP-bound $p21^{ras}$, the active state of $p21^{ras}$ (data not shown).

The results presented so far indicates that stretch activates multiple second messenger systems simultaneously. The activation of multiple second messenger system is reminiscent to humoral growth factor response (Cantley *et al.*, 1991; Rosengurt, 1991). To examine the possibility that stretch may cause release of some growth factor(s), we examined effects of 'stretch-conditioned' media on non-stretched cells. Cardiocytes were stretched for short periods and the stretch-conditioned media were transferred to non-stretched cardiocytes. Although transferring the media from the non-stretched cardiocytes did not elicit *c-fos* response, the media conditioned by 1, 5 and 10 min of stretch caused 3-, 9- and 10-fold increases in *c-fos* expression in non-stretched cardiocytes, respectively. The results indicate that stretch may cause a release of factor(s) into the culture media, which in turn induces *c-fos* expression. Stretch-conditioned media also activated MAP kinases in recipient cells (data not shown).

DISCUSSION

In order to elucidate how mechanical forces are transduced into intracellular signals regulating gene expression, we have investigated the roles of various signal transduction pathways in stretch-induced *c-fos* gene expression in rat cardiac myocytes using an *in vitro* model of load-induced cardiac hypertrophy. Figure 4 depicts a hypothetical signal transduction cascade that incorporates most of the available data (Sadoshima *et al.*, 1992a, 1992b, 1993), although the order of activation and proposed interactions with other signaling molecules are still conjectural (see legend). At present we do not know what is the initial mechanotransducer ("stretch receptor") in this cascade. A potential candidate is altered interaction between extracellular matrix (ECM) and a putative stretch-receptor (ST-R) that may be associated with a tyrosine kinase (TK) as depicted in the figure, although we have no direct evidence for this hypothesis. Within one minute of stretch, several proteins are tyrosine-phosphorylated, one of which is MAP kinase (MAPK). MAPK activates one of ribosomal S6 kinases (RSK) which phosphorylates the serum response factor (SRF) (Blenis, 1991). MAPK can also phosphorylate $p62^{TCF}$ to promote a ternary complex formation with SRF on the serum response element (SRE) of the *c-fos* gene (Gille et al., 1992). $p21^{ras}$ is also activated by stretch but its mechanism of activation is unknown. A biochemical link between PKC and MAPK activator (MAPKA) is not clear in cardiac myocytes but recent studies in PC12 cells suggest that $p21^{ras}$ activation is required for PKC-induced activation of MAPK (Woods et al., 1992).

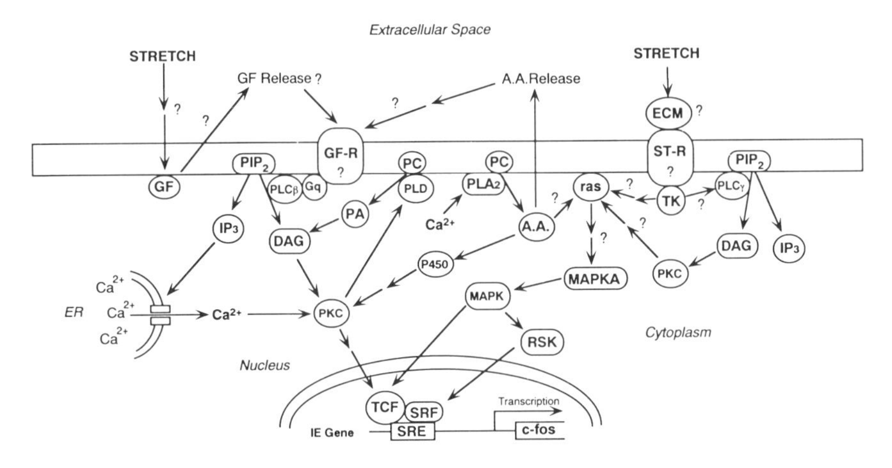

FIGURE 4 . A hypothetical signal transduction pathway of stretch-induced *c-fos* expression.
The arrows provide the notion of directionality in the signaling processes, some of which are still conjectural (marked with ?) or modeled after the known hierarchy of activation in other cell systems. Double arrows indicate unknown numbers of steps between the two signaling molecules.

A soluble growth factor (GF), whose molecular identity remains to be determined, may be released within a few minutes of stretch. This presumably activates its receptor (GF-R) which may couple to G protein (as depicted in Figure 4) or may couple to a receptor tyrosine kinase. Measurements of second messengers have indicated that stretch causes generation of inositol trisphosphate (IP_3), diacyl glycerol (DAG), phosphatidic acid (PA) and arachidonic acids (AA) by hydrolysis of phosphatidylinositol bisphosphate (PIP_2) and phosphatidylcholine (PC). DAG causes activation of protein kinase C (PKC) which may lead to activation of phospholipase D (PLD) (Exton, 1990), resulting in a sustained elevation of DAG and PKC activity as we observed. IP_3 can be generated either by phospholipase C_β (PLC_β) coupled to G protein or by PLC_γ activated by TK (Cantley et al., 1991). Although we did not measure intracellular Ca^{2+}, generation of IP_3 is expected to cause Ca^{2+} release from internal stores (ER). Released Ca^{2+} may activate PKC and phospholipase A_2 (PLA_2). Stretch causes a release of AA from the cell and some AA metabolites may potentially act as a paracrine/antocrine factor, while others may be metabolized by cytochrome P450 monoxygenase (P450) to stimulate PKC activity (Nishizuka, 1992; Piomelli and Greengard, 1990).

In summary, the stretch response causes a rapid activation of multiple second messenger systems, including tyrosine kinases, $p21^{ras}$, MAP kinases, S6 kinases, PLC, PLD, PKC and arachidonic acid release, many of which seem to contribute to *c-fos* gene induction by stretch. On the other hand, the cAMP and Ca^{2+}/CaM kinase pathways do not seem to be essential for stretch response (Sadoshima and Izumo, 1993), although we cannot rule out a potential permissive role of Ca^{2+}. The stretch-induced signal transduction pathways of the *c-fos* gene transcription are likely to converge into activation of $p62^{TCF}$-SRF complex on SRE. Other signals generated by stretch may also participate in a complex growth (hypertrophic) response of cardiac myocytes. It is intriguing that the activation of multiple signal transduction pathways by stretch is highly reminiscent of the complex cellular response to growth factors (Rozengurt, 1991). In fact, the stretch response seems to involve an autocrine/paracrine mechanism, because the stretch-conditioned medium mimics the effect of stretch. In retrospect, it is not surprising that the stretch response resembles the growth factor response of quiescent cells, because the hypertrophic response of cardiac myocytes in response to mechanical overload *in vivo* and *in vitro* closely resembles the mitogenic response of other cell types to growth factor stimulation (Izumo *et al.*, 1988; Sadoshima *et al.*, 1992a). It remains to be determined what is (are) the factor(s) released in response to stretch and how mechanical stretch leads to release of such factor(s).

ACKNOWLEDGEMENT

We thank Drs. M. Gilman and J. Blenis for plasmids and antibodies. This research was supported in part by the Jacob D. Indursky Memorial Fund to Beth Israel Hospital. J.S. is a Fellow of the Human Frontier Science Program and S.I. is an Established Investigator of the American Heart Association. We thank Jon Goldman for preparing this manuscript.

REFERENCES

Berridge MJ, Heslop JP, Irvine RF, and Brown KD (1984): Inositol trisphosphate formation and calcium mobilization in Swiss 3T3 cells in response to platelet derived growth factor. *Biochem J* 222: 195-201.

Blenis J. (1991): Growth regulated signal transduction by the MAP kinases and RSKs. *Cancer Res.* 3:445-449.

Cantley LC, Auger KR, Carpenter C, Duckworth B, Graziani A, Kapeller R, and Soltoff S. (1991): Oncogenes and signal transduction. *Cell* 64:281-302.

Exton JH (1990): Signaling through phosphatidylcholine breakdown. *J Biol Chem* 265: 1-4.

Gille H, Sharrocks AD, and Shaw PE (1992): Phosphorylation of transcription factor $p62^{TCF}$ by MAP kinase stimulates ternary comlex formation at c-fos promoter. *Nature* 358:414-417.

Graham R, and Gilman M (1991): Distinct protein targets for signals acting at the *c-fos* serum response element. *Science* 251: 189-192.

Ingber D (1991): Integrins as mechanochemical transducers. *Curr Opin in Cell Biol* 3: 841-848.

Izumo S, Nadal-Ginard B, and Mahdavi V (1988): Protooncogene induction and reprogramming of cardiac gene expression produced by pressure overload. *Proc. Natl. Acad. Sci. USA* 85:339-343.

Komuro I, Kaida T, Shibasaki Y, Kurabayashi M, Katoh Y, Hoh E, Takaku F, and Yazaki Y (1990): Stretching cardiac myocytes stimulates protooncogene expression. *J. Biol. Chem.* 265:3595-3598.

Komuro I, Katoh Y, Kaida T, Shibazaki Y, Kurabayashi M, Hoh E, Takaku F, and Yazaki Y (1991): Mechanical loading stimulates cell hypertrophy and specific gene expression in cultured rat cardiac myocytes. *J Biol Chem* 266: 1265-1268.

Morgan HE, and Baker KM (1991): Cardiac Hypertrophy: Mechanical, neural, and endocrine dDependence. *Circulation* 83: 13-25.

Nishizuka Y (1992): Intracellular signaling by hydrolysis of phospholipids and activation of protein kinase C. *Science* 258;607-614.

Piomelli D, and Greengard P (1990): Lipoxygenase metabolites of arachidonic acid in neuronal transmembrane signalling. *Trends in Pharm Sci* 11: 367-373.

Rozengurt E (1991): Neuropeptides as cellular growth factors: role of multiple signalling pathways. *Europ J Clin Invest* 21: 123-134.

Sadoshima J, Jahn L, Takahashi T, Kulik TJ, and Izumo S (1992a): Molecular characterization of the stretch-induced adaptation of cultured cardiac cells. *J Biol Chem* 267:10551-10560.

Sadoshima J, Takahashi T, Jahn L, ands Izumo S (1992b): Roles of mechano-sensitive ion channels, cytoskeleton and contractile activity in stretch-induced immediate-early gene expression and hypertrophy of cultured cardiac myocytes. *Proc Natl Acad Sci USA* 89:9905-9909

Sadoshima J and Izumo S (1993): Mechanical stretch rapidly activates multiple signal transduction pathways in cardiac myocytes: potential involvement of autocrine/paracrine mechanism. *EMBO J*, in press

Satoh T, Nakafuku M, and Kaziro Y (1992): Function of *ras* as a molecular switch in signal stransduction. *J. Biol. Chem.* 267:24149-24152.

Schalasta G, and Doppler C (1990): Inhibition of *c-fos* transcription and phosphorylation of the serum response factor by an inhibitor of phospholipase C-type reactions. *Mol. Cell. Biol.* 10:5558-5561.

Sheng M, and Greenberg ME (1990): The regulation and function of *c-fos* and other immediate early genes in the nervous system. *Neuron* 4: 477-485.

Vandenburgh HH, and Kaufman S (1979): *In vitro* model for stretch-induced hypertrophy of skeletal muscle. *Science* 203: 265-267.

Vandenburgh HH (1992): Mechanical forces and their second messengers in stimulating cell growth in vitro.*Am J Physiol* 262: R350-R355.

Wood KW, Sarnecki C, Roberts TM, and Blenis J (1992): *ras* mediates nerve growth factor receptor modulation of three signal-transducing protein kinases: MAP kinase, Raf-1 and RSK. *Cell* 68:1041-1050.

SHEAR STRESS-INDUCED GENE EXPRESSION IN HUMAN ENDOTHELIAL CELLS

Hsyue-Jen Hsieh, Nan-Qian Li, and John A. Frangos*

Department of Chemical Engineering, Pennsylvania State University, University Park, PA 16802

* To whom correspondence should be addressed.

INTRODUCTION

Cells are surrounded by an extracellular fluid which contains nutrients essential for maintenance of cellular life. The transport of nutrients therefore significantly influences the metabolic function of cells. A single prokaryotic cell can readily obtain enough nutrients from the environment via diffusional transport. For large, multicellular organisms, such as humans, transport by diffusion alone, however, is unable to deliver sufficient nutrients to and remove metabolic wastes from cells. A circulatory system thus exists to provide rapid, convective mass transport throughout the body, supplemented by the diffusional transport between the blood capillaries and the cells via extracellular fluid. In vertebrates the circulatory system supplies the body with nutrients and removes metabolites via blood flow. It is therefore indispensable in maintaining a normal physiological state. Dysfunction of the circulatory system may result in many diseases such as atherosclerosis, heart failure, and hypertension.

Atherosclerosis is the vascular disease responsible for more than 50% of the mortality in the developed countries. The development of atherosclerotic plaques by gradual accumulation of cells and extracellular matrix in the arterial walls can cause the luminal narrowing or occlusion of large arteries. Clinical evidence indicates that atherosclerotic lesions on human vessel walls do not develop randomly but instead localize at certain sites where the blood flow is disturbed and separation of streamlines and formations of eddies are likely to occur, i.e., regions of predicted low wall shear stress (Nerem and Cornhill, 1980; Davies, 1988). Therefore, hemodynamics in the arterial system seems to play an important role in the progression of atherosclerosis.

Vascular endothelial cells, which line the inner surface of blood vessels and are in direct contact with blood flow, appear to be essential in maintaining a normal hemodynamic environment. Wall shear stress applied on the endothelial cells by the blood flow plays a key role in mediating hemodynamic-endothelium interactions. The purpose of this report is to provide evidence that shear stress can regulate the endothelial functions by stimulating the expression of specific genes which are of physiological and pathological importance. A comprehensive study on the signalling pathway of shear-induced gene expression in endothelial cells will enhance the understanding of hemodynamics-endothelium interactions and eventually may lead to the discovery of disease-causing factors.

EFFECTS OF SHEAR STRESS ON PLATELET-DERIVED GROWTH FACTOR GENE EXPRESSION

The average wall shear stress in arterial vessels is maintained at 15 to 20 dynes/cm^2 (Kamiya and Togawa, 1980) by an autoregulatory mechanism that consists of two levels: acutely through vessel constriction or dilation; and chronically through the permanent change of arterial diameter. Vascular endothelium appears to be essential in autoregulating the wall shear stress (Langille and O'Donnell, 1986). The changes in the blood vessel size, however, can be achieved only by vascular smooth muscle cells (SMC), the major cells in the intima of the arterial wall. Therefore one of the roles of the endothelium is to sense blood flow and mediate any necessary response of SMC. Platelet-derived-growth factor (PDGF), a mitogen for SMC and also a vasoconstrictor, may be involved in the autoregulation of vascular wall shear stress.

Endothelial cells secrete PDGF that almost exclusively goes into the basal compartment (Zerwes and Risau, 1987). It is thus generally believed that PDGF released from vascular endothelial cells may bind to PDGF receptors on the underlying SMC *in vivo* (Ross et al., 1986b). PDGF genes are located on human chromosomes 7 and 22, encoding A-chain and B-chain polypeptides respectively (Betsholtz et al., 1986; Swan et al., 1982), with amino acid sequence homology of about 56% (Betsholtz et al., 1986). PDGF B-chain precursor is encoded by the c-sis gene, which is homologous to the transforming gene v-*sis* of simian sarcoma virus (SSV) (Stiles, 1983). Active PDGF is a dimer of two chains linked by disulfide bonds. Three possible forms of PDGF, i.e. A-A, A-B, and B-B dimers, have been identified (Ross et al., 1986b). Four PDGF-A transcripts (Starksen et al., 1987) and one PDGF-B transcript (Collins et al., 1987) were found in endothelial cells.

Because of its special functions in controlling the proliferation and migration of SMC, the major cells in the arterial wall, PDGF may play a key role in vascular diseases such as atherosclerosis. Elevated PDGF-B transcript levels in human atherosclerotic plaques compared to normal artery have been reported (Barrett and Benditt, 1987). Nevertheless, in normal bovine endothelial cells *in vivo*, a persistent low level PDGF-B transcription was observed, suggesting that PDGF gene may also have a persisting role in maintenance of vascular wall (Barrett et al., 1984).

To examine the possible role of PDGF in the hemodynamic response of vascular wall, cultured human umbilical vein endothelial cells (HUVEC) were subjected to steady shear stress in a parallel plate flow chamber (Frangos et al., 1988). The levels of both PDGF A-chain and B-chain mRNA in HUVEC were increased by a physiological shear stress (16 dynes/cm^2), reaching a maximum about 1.5 to 2 hours after the onset of shear stress and returning almost to basal levels at 4 hours (Figures 1A and 1B) (Hsieh et al., 1991). Glyceraldehyde-3-phosphate dehydrogenase (GAPDH) mRNA levels, used as an internal control, were relatively constant (Figure 1C). The peak levels showed a more than 10-fold enhancement for PDGF-A mRNA and a 2 to 3-fold increase for PDGF-B mRNA (Figures 1D and 1E). PDGF-A mRNA also showed a shear stress intensity-dependent increase from 0 to 6 dynes/cm^2, and then plateaued from 6 to 51 dynes/cm^2 (Figure 2A) (Hsieh et al., 1991). PDGF-B mRNA levels were increased as shear stress increased from 0 to 6 dynes/cm^2, then declined gradually to a minimum at 31 dynes/cm^2, and increased again when shear stress rose to 51 dynes/cm^2 (Figure 2B).

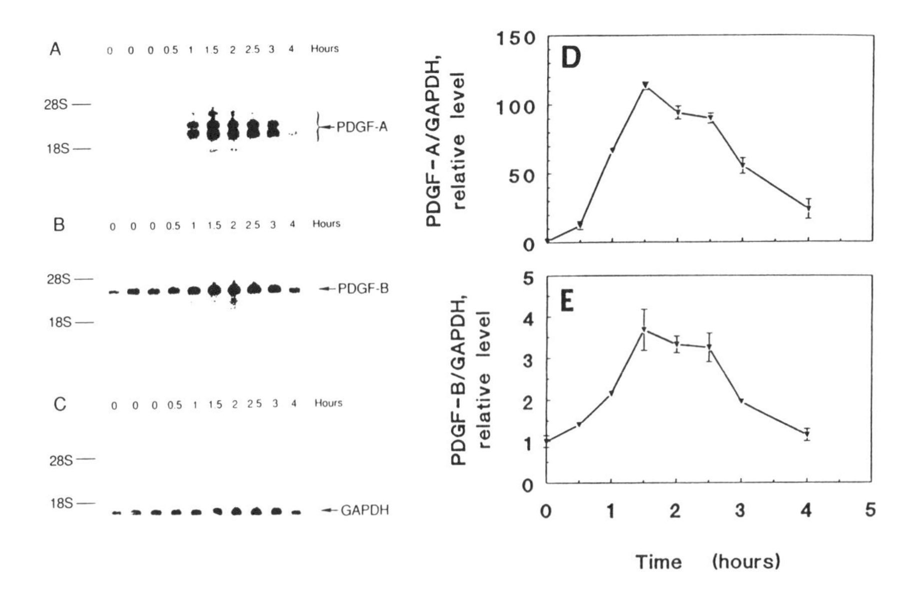

FIGURE 1. Time course of PDGF-A and PDGF-B mRNA induction by shear stress. HUVEC were subjected to steady shear stress of 16 dynes/cm^2 for 0.5, 1, 1.5, 2, 2.5, 3, and 4 hrs. Total RNA was isolated from cells for Northern analysis of PDGF-A (panel A), PDGF-B (panel B), and GAPDH mRNA (panel C). PDGF mRNAs were induced after 1.5 h of steady flow. GAPDH mRNA levels were relatively constant. For comparison, 18S and 28S rRNA markers are shown. The intensity of each hybridization band on autoradiograms was determined by optical densitometry. Panels D and E show the time course of relative PDGF-A and PDGF-B mRNA levels, respectively (normalized by GAPDH mRNA). Each data point represents the mean of three experiments ± SEM. (n=11 for stationary controls (0 hr)). Reprinted with permission from Hsieh et al. (1991).

The results indicate that the expression levels of PDGF-B are at their lowest in the range of 15 to 31 dynes/cm^2 (in cells subjected to shear), suggesting that PDGF, a mitogen for SMC (Ross et al., 1986b) and a potent vasoconstrictor (Berk et al., 1986), may be involved in the autoregulation of wall shear stress according to the following hypothesis. If the decrease in PDGF-B expression levels with increasing shear (in the range of 6 to 31 dynes/cm^2) translates into decreased PDGF protein secretion, a vasodilatory (decreased vasoconstrictory) signal will result (Figure 2B). The role of PDGF in flow-dependent vasomotor control would be in conjunction with other known mediators of flow-dependent dilation, such as endothelial-derived relaxing factor (EDRF) and prostacyclin. Chronically, altered PDGF secretion may regulate the chronic changes in vascular caliber in the process of autoregulating wall shear stress. The role of flow-dependent PDGF secretion is not so clear in this case. A reduction in vascular wall shear stress leads to a reduction in vessel diameter which is not due to growth of vessel wall tissue but rather to a change in vessel size (Langille and O'Donnell, 1986). In contrast, shear stress of 51 dynes/cm^2 that is higher than the average arterial shear stress (15 to 31 dynes/cm^2) also induced increased expression of PDGF-B (Figure 2B). This may partly explain how increased blood flow leads to increased media cross-sectional area (Zarins et al., 1987). It is likely, however, that flow-induced vascular remodeling involves several other interacting factors.

The increased expression of PDGF-B with decreasing shear established in this study may also be involved in atherogenic mechanisms. Intimal thickening or proliferation has been found in certain regions of low shear stress in blood vessels (Zarin et al., 1987), which may coincide with areas of vascular lesions, as observed by Caro et al. (1971). On the other hand, the elevated PDGF-B mRNA levels induced by high shear stresses (39 to 51 dynes/cm^2) (Figure 2B) seem consistent with the "response-to-injury hypothesis of atherogenesis" proposed by Ross (1986a). High shear stress may be seen as an injury to the endothelium, as proposed earlier by Fry, which may stimulate endothelial cells to secrete growth factors including PDGF, possibly involved in the atherogenesis. Great care, however, must be taken in extrapolating *in vitro* data to *in vivo* pathology.

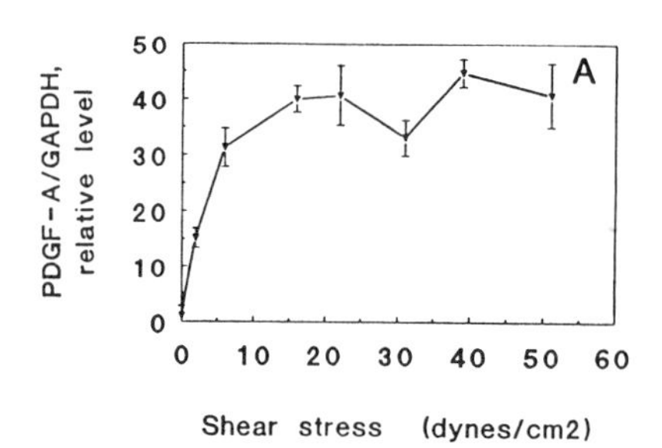

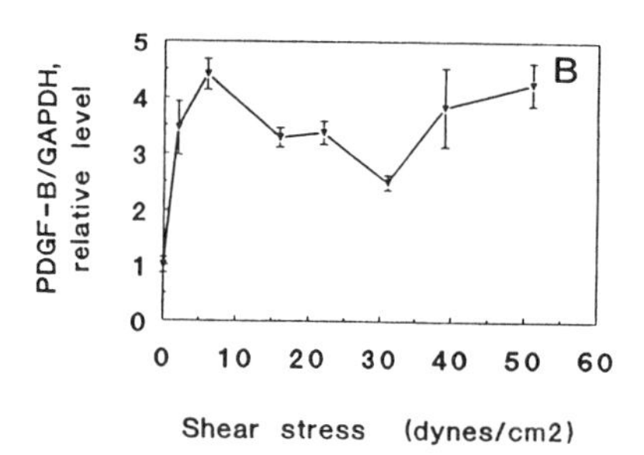

FIGURE 2. Effect of shear stress intensity on PDGF-A and PDGF-B mRNA levels. HUVEC were subjected to various levels of steady shear stress for 1.5 h. The intensities of PDGF-A and PDGF-B mRNA signals were normalized by GAPDH mRNA. PDGF-A mRNA levels (panel A) were higher at 6 dynes/cm^2 than those at 2 dynes/cm^2 ($p < 0.05$) and the mRNA levels at 2 dynes/cm^2 were higher than those at 0 dyne/cm^2 ($p < 0.05$). But no significant shear-dependence of PDGF-A mRNA levels was observed in the range of 6 to 51 dynes/cm^2. In contrast, the levels of PDGF-B mRNA (panel B) appeared to be varied within the range of 0 to 51 dynes/cm^2. In the low shear region, there was an increase of PDGF-B mRNA levels when shear stress was increased from 0 to 6 dynes/cm^2. However, PDGF-B mRNA levels at 6 dynes/cm^2 were significantly higher than those at 16 and 22 dynes/cm^2 ($p < 0.05$). PDGF-B mRNA levels at 16 and 22 dynes/cm^2 were significantly higher than those at 31 dynes/cm^2 ($p < 0.05$). In addition, PDGF-B mRNA levels at 51 dynes/cm^2 were also significantly higher than those at 31 dynes/cm^2 ($p < 0.05$). Data points represent the mean of several experiments ± SEM; n=16, 10, 5, 13, 8, 7, 5, and 7 for the 0, 2, 6, 16, 22, 31, 39, and 51 dynes/cm^2 shear stress points, respectively. Reprinted with permission from Hsieh et al. (1991).

EFFECTS OF SHEAR STRESS ON NUCLEAR PROTO-ONCOGENES c-*fos*, c-*jun*, and c-*myc* EXPRESSION

Nuclear proto-oncogenes c-*fos* and c-*jun* encode proteins belonging to the family of transcription factor AP-1, either as Jun-Fos heterodimers or Jun-Jun homodimers. c-Myc protein is a putative transcription factor or regulator. These proto-oncogene are called "immediate early genes", since they respond to various stimuli rapidly and play key roles in regulating the expression of many target genes (Seuwen and Pouyssegur, 1992). It has been established that shear stress increases the expression of tissue plasminogen activator (tPA) (Diamond et al., 1990) and PDGF (Hsieh et al., 1991) genes in HUVEC. It is thus likely that shear stress may also affect the expression of c-*fos*, c-*jun*, and c-*myc*. In this report, the effects of shear stress on these nuclear proto-oncogenes were examined.

c-*fos* mRNA levels in stationary cultures were very low. A 1 Hz pulsatile flow with an average shear stress of 16 dynes/cm^2 induced a dramatic increase of c-*fos* mRNA levels in HUVEC 0.5 h after the onset of flow, which declined rapidly to basal values within 1 h. Steady flow with a similar shear stress also induced a transient increase of c-*fos* mRNA levels, but to a lesser extent (Figure 3) (Hsieh et al., 1992b). In addition, increased c-*fos* mRNA levels were observed when low shear (2 to 6 dynes/cm^2) was replaced by high shear (16 to 33 dynes/cm^2). Pulsatile and steady flow caused a slight increase of c-*jun* and c-*myc* mRNA levels. The role of pulsatility was also investigated in PDGF expression. Pulsatile flow induced a transient increase of PDGF-A and PDGF-B mRNA levels with peaks at 1.5 to 2 h. Pulsatile flow, which was more stimulatory in mediating c-*fos* expression, however, was less stimulatory than steady flow in mediating PDGF expression.

A consensus DNA binding site for AP-1 has been identified in negative and positive regulatory regions of several genes (Cooper, 1990). Therefore, shear-induced expression of c-*fos*, in cooperation with c-*jun*, may play an important role in mediating many long-term cellular responses to shear stress, such as gene induction or regulation. Furthermore, Jun-Fos heterodimers showed notably higher DNA binding affinity than Jun-Jun homodimers (Abate et al., 1990). The dramatic increase in c-*fos* expression combined with the small increase of c-*jun* expression suggests a change in the DNA binding affinity as well as the total quantity of AP-1 in HUVEC exposed to fluid flow.

Increased expression of c-*fos*, c-*jun*, and c-*myc* in cells has been correlated to cell proliferationor differentiation (Cooper, 1990; DePinho et al., 1991). Therefore, the observed increase in the expression of these proto-oncogenes is somewhat surprising since others have shown that fluid flow inhibits the proliferation of confluent endothelial cells (Mitsumata et al., 1991). The role of these proto-oncogenes in flow stimulation of endothelial cells is therefore unclear, but it is quite likely that they are involved in cellular responses other than cell growth. It has been recently reported that c-Myc protein can bind *in vitro* to the sequence CACGTG (Blackwell et al., 1990). Recent studies suggest that c-Myc protein can associate with other proteins to form a "putative" transcription factor (Torres et al., 1992). In this study it has been found that there is a slight increase of c-myc transcription by shear stress, suggesting that an altered nuclear activity may exist in HUVEC.

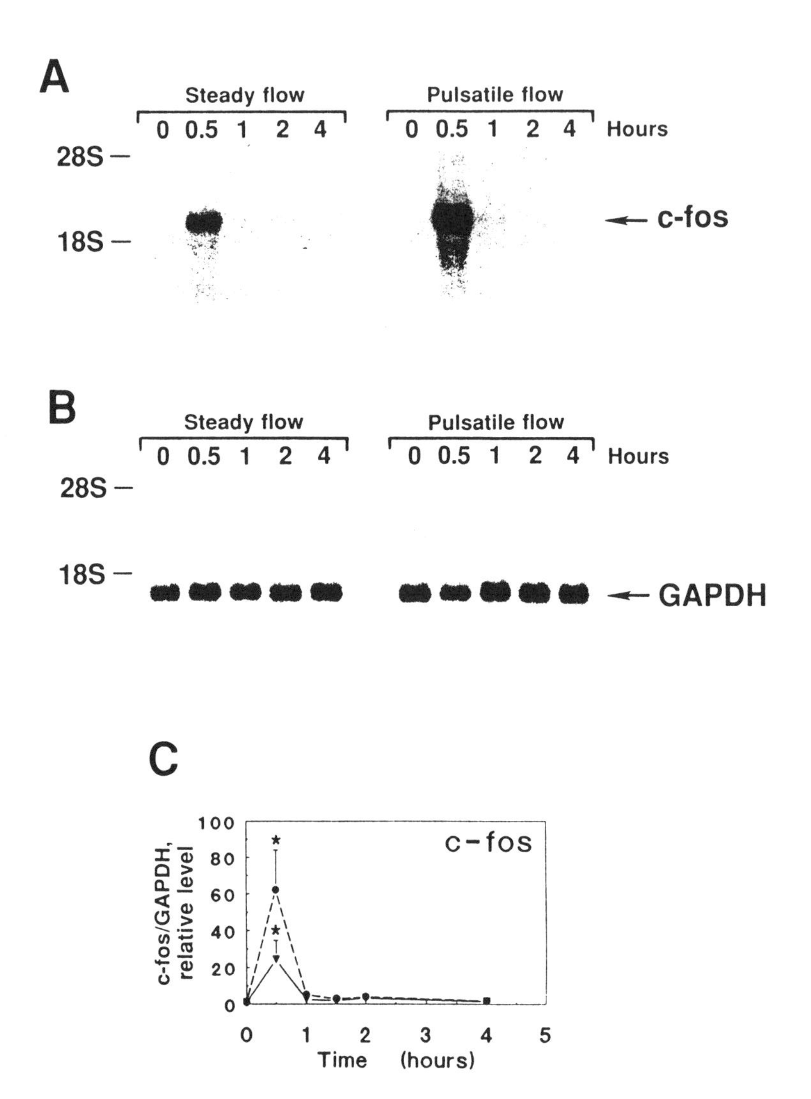

FIGURE 3. Time course of c-*fos* mRNA induction by pulsatile or steady flow. HUVEC were subjected to pulsatile (1 Hz) or steady flow with an average shear stress of 16 dynes/cm^2 for 0.5, 1, 2, and 4 hours. Total RNA was isolated from cells for Northern blot analysis of c-*fos* (panel A) and GAPDH mRNA (panel B). c-*fos* mRNA was induced after 0.5 h of pulsatile or steady flow. GAPDH mRNA levels were relatively constant. Panel C shows the time course of relative c-*fos* mRNA levels (normalized by GAPDH mRNA) under pulsatile (● – – – ●) or steady (▼——▼) flow condition. Data points represent the means of several experiments ± SEM; n=4, 3, 3, 3, 3, and 3 for the 0, 0.5, 1, 1.5, 2, and 4 h time points, respectively. * $P < 0.05$ vs. stationary controls. Reprinted with permission from Hsieh et al. (1992b)

Our results show that pulsatile flow was more stimulatory (3-fold) than similar levels of steady flow in inducing c-*fos* expression. This is consistent with previous observations that flow-induced prostacyclin production is significantly enhanced (2- to 3-fold) by pulsatility (Frangos et al., 1985). PDGF expression, however, did not respond similarly to pulsatile flow. The increase in PDGF expression by pulsatile flow was only one half of that by steady flow. On the other hand, it has been reported that pulsatile and steady flow both stimulate tPA secretion in HUVEC to the similar extent (Diamond et al., 1989). The mechanism by which unsteadiness in flow modulates shear-induced responses is as of yet unclear. These findings indicate the existence of a complex, flow-mediated regulatory mechanism of gene expression in HUVEC.

MECHANISM OF SHEAR-INDUCED GENE EXPRESSION

The mechanisms for the shear-induced PDGF gene expression are not well understood. Previous studies have shown that shear stress can stimulate membrane phosphoinositide turnover (perhaps via phospholipase C) in endothelial cells, producing the second messengers inositol trisphosphate (IP_3) and diacylglycerol (DAG) (Bhagyalakshmi et al., 1992; Nollert et al., 1990). IP_3 triggers the release of calcium from intracellular pools, and DAG is an activator of protein kinase C, a major kinase responsible for many cellular responses, e.g., gene expression, cell proliferation, etc. (Nishizuka, 1986). Moreover, the release of PDGF from endothelial cells has been reported to be greatly stimulated by thrombin (Harlan et al., 1986) and PMA (Starksen et al., 1987). Thrombin leads to inositol phospholipid turnover. PMA, like diacylglycerol, is an activator of protein kinase C. Therefore, shear-induced PDGF expression appears consistent with a signal transduction mechanism involving protein kinase C activation, mediated by second messenger IP_3 and diacylglycerol.

Increase by shear of intracellular levels of second messengers cAMP and cGMP recently have been reported (Reich et al., 1990; Kuchan and Frangos, 1991). The effects of cAMP or cGMP are assumed to be mediated through activation of cAMP- or cGMP-dependent protein kinases. cAMP-dependent kinase seems to be responsible for regulating cAMP-dependent gene expression, presumably via the transcription factor cAMP response element-binding protein (CREB) that binds to the cAMP response element (CRE) in the regulatory region of a gene (Karin, 1991). As previously mentioned, G proteins are important regulatory proteins in the intracellular signal transduction pathways. In agonist-stimulated endothelial cells, G proteins have been demonstrated to be responsible for the regulation of phospholipase C (Voyno-Yasenetskaya et al., 1989). Recently, a pertussis toxin-sensitive G protein has been found to be responsible for mediating flow-induced prostacyclin production in HUVEC (Berthiaume and Frangos, 1992). Intracellular calcium is a second messenger involved in many cellular events. It may act as an activator of several kinases including protein kinase C (Nishizuka, 1986).

By using various inhibitors, the signal transduction pathway of shear-induced PDGF and c-*fos* expression in HUVEC was elucidated. The results suggest that shear-induced PDGF gene expression in HUVEC is mainly mediated by protein kinase C with the involvement of G proteins and intracellular calcium (Hsieh et al., 1992a). Subsequent studies indicate that protein kinase C is also an important mediator in flow-induced c-*fos* expression (Hsieh et al., 1992b), suggesting a possible link between PDGF and c-*fos* induction.

ROLES OF c-*fos* IN SHEAR-INDUCED CELLULAR RESPONSES

One of the most likely roles of c-*fos*, as a transcription factor, is to mediate shear-induced gene expression such as PDGF and tPA induction. To examine this hypothesis, a c-*fos* antisense oligodeoxynucleotide was utilized to block the expression of c-*fos* gene. The results indicated that c-*fos* antisense significantly inhibited shear-induced c-*fos* transcription, but failed to block shear-induced PDGF and tPA expression, suggesting that c-*fos* induction by shear may not play an important role in the mechanism of shear-induced PDGF and tPA expression. It has been recently reported that flow-induced PDGF-B expression in bovine aortic endothelial cells is possibly mediated by novel transcription factors which are different from AP-1 (Resnick et al., 1992). Their findings therefore are consistent with our results obtained from c-*fos* antisense experiments.

The immunofluorescence staining of c-Fos protein in HUVEC demonstrate that even in the stationary confluent cultures there is a basal level of c-Fos protein which can be slightly increased by subjecting cells to shear stress, presumably due to the shear-induced expression of c-*fos* gene. The existence of c-Fos protein in stationary HUVEC is unexpected because in most quiescent cells c-Fos protein is undetectable (Distel and Spiegelman, 1990). Nevertheless, considering that HUVEC *in vivo* are constantly sheared by the blood flow, HUVEC in the stationary culture condition are therefore not in their normal environment. Thus stationary HUVEC even reaching confluence are probably not quiescent cells. c-Fos protein may play a role in several phenomena observed in the stationary HUVEC such as elevated transcription and production rate of endothelin compared to sheared HUVEC (Sharefkin et al., 1991).

The existence of c-Fos protein downregulates the expression of c-*fos* gene (Sasson-Corsi et al., 1988), which may help to explain why there is a dramatic increase in c-*fos* transcription by shear stress but not in translation. The change of c-Fos protein activity by phosphorylation has been demonstrated in the literature (Barber and Verma, 1987). Even though the immunofluorescence study indicated only a slight increase of c-Fos protein levels in the shear-stimulated cells; it is possible that shear stress alters its activity by changing the phosphorylation state of the protein activity.

CONCLUSIONS

Mechanical forces such as pressure, stretch, and fluid shear stress have been previously demonstrated to regulate the metabolic functions of cells. These phenomena may play important physiological roles *in vivo*. In the case of vascular endothelial cells, shear stress mediates many of the adaptive responses of blood vessels to flow, such as altered production rates for vasodilatory and vasoconstrictory compounds. Understanding the regulatory role of mechanical forces on the function of cells is an important area in tissue engineering.

The present study reveals that fluid shear stress induces PDGF and c-*fos* expression in HUVEC in a shear intensity-dependent manner, implying that they may be involved in the adjustment of blood vessels to flow. Shear stress also increases c-*jun* and c-*myc* mRNA levels slightly. PDGF and c-*fos* induction by shear is probably mediated by protein kinase C, with the involvement of G proteins, phospholipase C, and intracellular calcium. Immunofluorescence studies indicate that in stationary HUVEC cultures there is a basal level of c-Fos protein, which can be slightly increased by shear stress. Pulsatile and steady shear stress induces gene expression to a different extent, indicating that a complex, shear-mediated regulatory mechanism of gene expression exists. The increased expression of c-*fos*, c-*jun*, and c-*myc* may regulate long-term cellular responses to shear stress.

ACKNOWLEDGEMENTS

We thank C. Betsholtz for providing the PDGF A chain cDNA plasmid, C. D. Rao for providing the PDGF B chain cDNA plasmid, and Upjohn Co. for providing the phospholipase C inhibitor U73122. We also thank F. Berthiaume and M. J. Kuchan for helpful discussion, and N.-W. Chow for valuable technical assistance. This investigation was supported by National Heart, Lung, and Blood Institute Grant HL-40696. JAF is a recipient of the National Science Foundation Presidential Young Investigator Award.

REFERENCES

Abate C, Luk D, Gentz R, Rauscher FJ, Curran T: Expression and purification of the leucine zipper and DNA-binding domains of Fos and Jun: Both Fos and Jun contact DNA directly. Proc Natl Acad Sci USA 1990;87:1032-1036

Barber J, Verma IM: Modification of Fos proteins: phosphorylation of c-fos, but not v-fos, is stimulated by 12-tetradecanoyl-phorbol-13-acetate and serum. Mol Cell Biol 1987;7:2201-2211

Barrett TB, Gajdusek CM, Schwartz SM, McDougall JK, Benditt EP: Expression of the sis gene by endothelial cells in culture and in vivo. Proc Natl Acad Sci USA 1984;81:6772-6774

Barrett TB, Benditt EP: sis (platelet-derived growth factor B chain) gene transcript levels are elevated in human atherosclerotic lesions compared to normal artery. Proc Natl Acad Sci 1987;84:1099-1103

Berk BC, Alexander RW, Brock TA, Gimbrone MA, Webb RC: Vasoconstriction: A new activity for platelet-derived growth factor. Science 1986;232:87-90

Berthiaume F, Frangos JA: Flow-induced prostacyclin production is mediated by a pertussis toxin-sensitive G protein. FEBS Lett 1992;(in press)

Betsholtz C, Johnsson A, Heldin C-H, Westermark B, Lind P, Urdea MS, Eddy R, Shows TB, Philpott K, Mellor AL, Knott TJ, Scott J: cDNA sequence and chromosomal localization of human platelet-derived growth factor A-chain and its expression in tumor cell lines. Nature 1986;320:695-699

Bhagyalakshmi A, Berthiaume F, Reich KM, Frangos JA: Fluid shear stress stimulates membrane phospholipid metabolism in cultured human endothelial cells. J Vasc Res 1992;(in press)

Blackwell TK, Kretzner L, Blackwood EM, Eisenman RN, Weintraub H: Sequence-specific DNA binding by the c-Myc protein. Science 1990;250:1149-1151

Caro CG, Fitz-Gerald JM, Schroter RC: Atheroma and arterial wall shear: Observation, correlation and proposal of a shear dependent mass transfer mechanism for atherogenesis. Proc Roy Soc Lond B 1971;177:109-159

Collins T, Pober JS, Gimbrone MA, Hammacher A, Betsholtz C, Westermark B, Heldin C-H: Cultured human endothelial cells express platelet-derived growth factor A chain. Am J Pathol 1987;127:7-12

Cooper GM: Oncogenes. Boston, Jones and Bartlett Publishers, 1990;pp 230-235

Davies PF: Endothelial cells, hemodynamic forces, and the localization of atherosclerosis, in Ryan US (ed): Endothelial Cells. Boca Raton, FL, CRC Press Inc, 1988, vol 2, pp 123-138

DePinho RA, Schreiber-Agus N, Alt FW: myc family oncogenes in the development of normal and neoplastic cells. Adv Cancer Res 1991;57:1-46

Diamond SL, Eskin SG, McIntire LV: Fluid flow stimulates tissue plasminogen activator secretion by cultured human endothelial cells. Science 1989;243:1483-1485

Diamond SL, Sharefkin JB, Diffenbach C, Frasier-Scott K, McIntire LV, Eskin SG: Tissue plasminogen activator messenger RNA levels increase in cultured human endothelial cells exposed to laminar shear stress. J Cell Physiol 1990;143:364-371

Distel RJ, Spiegelman BM: Protooncogene c-fos as a transcription factor. Adv Cancer Res 1990;55:37-55.

Frangos JA, Eskin SG, McIntire LV, Ives CL: Flow effects on prostacyclin production by cultured human endothelial cells. Science 1985;227:1477-1479

Frangos JA, McIntire LV, Eskin SG: Shear stress induced stimulation of mammalian cell metabolism. Biotechnol Bioeng 1988;32:1053-1060

Harlan JM, Thompson PJ, Ross RR, Bowen-Pope DF: a-thrombin induces release of platelet-derived growth factor-like molecule(s) by cultured human endothelial cells. J Cell Biol 1986;103:1129-1133

Hsieh HJ, Li NQ, Frangos JA: Shear stress increases endothelial platelet-derived growth factor mRNA levels. Am J Physiol 1991;260:H642-H646

Hsieh HJ, Li NQ, Frangos JA: Shear-induced platelet-derived growth factor gene expression in human endothelial cells is mediated by protein kinase C. J Cell Physiol 1992a;150:552-558

Hsieh HJ, Li NQ, Frangos JA: Pulsatile and steady flow induces c-fos expression in human endothelial cells. J Cell Physiol 1992b;(in press)

Kamiya A, Togawa T: Adaptive regulation of wall shear stress to flow change in the canine carotid artery. Am J Physiol 1980;239:H14-H21

Karin M: Signal transduction and gene control. Curr Opin Cell Biol 1991;3:467-473

Kuchan MJ, Frangos JA: Influence of steady shear stress on cGMP concentrations in human umbilical vein endothelial cells, in: Mechanical Stress Effects on Vascular Cells, a workshop sponsored by Emory University, Georgia Institute of Technology, and Emory-Georgia Tech Biomedical Technology Research Center, Atlanta, GA, 1991, p.8

Langille BL, O'Donnell F: Reductions in arterial diameter produced by chronic decreases in blood flow are endothelium-dependent. Science 1986;231:405-407

Mitsumata M, Nerem RM, Alexander RW, Berk B: Shear stress inhibits endothelial cell proliferation by growth arrest in the G0/G1 phase of the cell cycle. FASEB J 1991;5:A527 (abstract #904)

Nerem RM, Cornhill JF: The role of fluid mechanics in atherogenesis. J Biomech Eng 1980;102:181-189

Nishizuka Y: Studies and perspectives of protein kinase C. Science 1986;233:305-312

Nollert MU, Eskin SG, McIntire LV: Shear stress increases inositol trisphosphate levels in human endothelial cells. Biochem Biophys Res Commun 1990;170:281-287

Reich KM, Gay CV, Frangos JA: Fluid shear stress as a mediator of osteoblast cyclic adenosine monophosphate production. J Cell Physiol 1990;143:100-104

Resnick N, Dewey CF, Atkinson W, Collins T, Gimbrone MA: Shear stress regulates endothelial PDGF-B chain expression via induction of novel transcription factors. FASEB J 1992;6:A1592 (Abstract: #3798)

Ross R: The pathogenesis of atherosclerosis - an update. New Engl J Med 1986a;314:488-500

Ross R, Raines EW, Bowen-Pope DF: The biology of platelet-derived growth factor. Cell 1986b;46:155-169

Sasson-Corsi P, Sisson JC, Verma IM: Transcriptional autoregulation of the proto-oncogene fos. Nature 1988;334:314-319

Seuwen K, Pouyssegur J: G protein-controlled signal transduction pathways and the regulation of cell proliferation. Adv Cancer Res 1992;58:75-94

Sharefkin JB, Diamond SL, Eskin SG, McIntire LV, Diffenbach CW: Fluid flow decreases preproendothelin mRNA levels and suppresses endothelin-1 peptide release in cultured human endothelial cells. J Vasc Surg 1991;14:1-9

Starksen NF, Harsh GR, Gibbs VC, Williams LT: Regulated expression of the platelet-derived growth factor A chain gene in microvascular endothelial cells. J Biol Chem 1987;262:14381-14384

Stiles CD: The molecular biology of platelet-derived growth factor. Cell 1983;33:653-655

Swan DC, McBride OW, Robbins KC, Keithley DA, Reddy EP, Aaronson SA: Chromosomal mapping of the simian sarcoma virus onc gene analogue in human cells. Proc Natl Acad Sci USA 1982;79:4691-4695

Torres R, Schteiber-Agus N, Morgenbesser SD, DePinho RA: Myc and Max: a putative transcriptional complex in search of a cellular target. Curr Opin Cell Biol 1992;4:468-474

Voyno-Yasenetskaya TA, Tkachuk VA, Cheknyova EG, Panchenko MP, Grigorian GY, Vavrek RJ, Stewart JM, Ryan US: Guanine nucleotide-dependent, pertussis toxin-insensitive regulation of phosphoinositide turnover by bradykinin in bovine pulmonary artery endothelial cells. FASEB J 1989;3:44-51

Zarins CK, Zatina MA, Giddens DP, Ku DN, Glagov S: Shear stress regulation of artery lumen diameter in experimental atherogenesis. J Vasc Surg 1987;5:413-420

Zerwes H-G, Risau W: Polarized secretion of a platelet-derived growth factor-like chemotactic factor by endothelial cells in vitro. J Cell Biol 1987;105:2037-2041

Section VI

MATERIALS FOR TISSUE REMODELING *IN VIVO* AND *IN VITRO*

TISSUE ENGINEERING OF SKELETAL AND CARDIAC MUSCLE FOR CORRECTION OF CONGENITAL AND GENETIC ABNORMALITIES AND RECONSTRUCTION FOLLOWING PHYSICAL DAMAGE.

GEOFFREY GOLDSPINK, Unit of Molecular and Cellular Biology, The Royal Veterinary College, University of London. Present address, Department of Anatomy and Developmental Biology, Royal Free Medical School, University of London, Rowland Hill Street, London NW3 2PF.

INTRODUCTION

Tissue engineering is a broad term which may embrace a variety of procedures ranging from physical manipulation to activate certain endogenous genes to the introduction of novel genes in order to alter the basic characteristics of the tissue. To appreciate the wide ranging possibilities it is perhaps appropriate to commence with some background information concerning the cellular and molecular responses of muscle to altered physical conditions.

Muscle has an inherent adaptability and a remarkable ability to adjust to the type of activity to which it is habitually subjected. Muscle adapts to a mechanical overload by undergoing hypertrophy and this has been shown to be the result of increase in size of the existing muscle fibres and not an increase in fibre number. The increase in girth of the existing fibres is associated with an increase in the number of myofibrils within each fibre which enables each fibre to produce more total force. The increased gene expression that is stimulated by a mechanical overload results in a greatly increased synthesis of contractile proteins such as actin and myosin. These are apparently added to the periphery of the existing myofibrils which grow to a certain size and then split longitudinally, into two or more daughter myofibrils. This splitting is due to a mismatch in the spacing of the actin and myosin lattices which causes a mechanical stress to develop in the centre of the Z disc when the myofibril reaches a certain critical size and is required to contract forcibly. Thus the myofibril mass is not built up as a complete intact mass but it is subdivided by the splitting mechanism. This enables the SR and T systems that are involved in the activation of the contractile system to invade the myofibril mass (Goldspink 1984).

In adaptation for fatigue resistance rather than increased power the muscle fibres undergo a phenotypic change in that the myofibrils are rebuilt using different "slow" type isoforms of the contractile proteins. Myosin cross bridges, which are the force generations for muscular contraction, are encoded in the myosin heavy chain (hc) genes. These exist as a family of individual genes, each one encoding for a different type of cross bridge which is expressed under certain specific conditions. This gene family includes an embryonic, a neonatal, an adult fast and an adult slow myosin hc gene. Other muscle genes such as the myosin light chain which appear to fine tune the rate of which the cross bridges work are encoded by single genes which are alternatively spliced to express the different isoforms. Our research has indicated that adult slow myosin hc genes are switched on by stretch and repetitive contractions (Goldspink *et al.,* 1992). When there is lack of stretch there only the fast genes are expressed (Loughna *et al.,* 1990). The new myosin replaces the existing myosin in the thick filaments by exchange in skeletal (Fischman *et al.,* 1986) and in cardiac muscle (Wenderoth and Eisenberg *et al.,* 1987) The resulting cross bridges have a long cycle time and split ATP at a slow rate and are thus much more economical for carrying out slow repetitive movements and generating isometric tension for the maintenance of posture (Goldspink 1977).

The subset of genes that is expressed when muscles adapt for increased fatigue resistance also includes energy metabolism genes that are part of the mitochondrial and nuclear genome (Williams *et al.,* 1986). The increased fatigue resistance also involves an increased rate of supply of ATP as well as the decreased rate of utilization by the slower myosin cross bridges. However, if a muscle is required for a high power output such as sprinting, the training regime must be of short enough generation not to turn on the slow genes but long enough to induce hypertrophy of the fibres. In this way the muscle remains fast but its mass increases, hence the total force and the rate at which the muscle can develop force are high. In this situation there is no way the supply of ATP can keep up with the rate its utilization. and mitochondrial density is not increased as the mitochondria would take up too much as the fibre mass at the expense of myofibrils.

As well as changing the phenotype and the cross sectional area of the fibres, muscle is also capable of adapting to a new functional length. When muscles are stretched during growth (Goldspink 1968), or altered posture (Tabary *et al.,* 1972), they add on sarcomeres serially to the ends of the existing myofibrils (Williams *et al.,* 1971). This has important physiological implications. Force development depends on the number of the myosin cross bridges that can interact with the actin filaments. There is therefore an optimum overlap of the myosin and actin filaments and hence an optimum sarcomere length. The only way of adjusting the sarcomere length is to adjust the number of sarcomeres in series. Muscle can therefore adjust to a changed functional length either by rapidly adding more sarcomeres in series or by taking off sarcomeres as the case may be. Indeed, dynamic stretch is very important in regulating protein synthesis and in maintaining muscle size (Loughna *et al.,* 1986). When the tendons are severed and the tissue is no longer under tension it undergoes atrophy in girth as well as length. The rate of tissue atrophy following tenotomy differs considerably between different types of muscle even when the nerve and blood supply remain intact (McMinn and Vrbova 1964). Postural muscles that are composed of slow oxidative skeletal muscle fibres show very rapid tissue loss. This must be borne in mind when translocating muscle flaps for grafting muscles and using translocated muscle flaps for heart assist or neosphincter construction.

As muscle genes respond to mechanical signals the prospects exist for engineering the tissue by altering the physical environment and thus altering the expression of endogenous genes. However, the possibility of introducing new genes using muscle stem cells and even by direct injection of DNA. Skeletal and cardiac muscle have been shown to have the unique ability of taking up and expressing DNA constructs following a single intramuscular injection (Wolff *et al.,* 1990). This offers the possibility of introducing engineered genes which include a coding sequence driven by a muscle specific promoter (regulators) sequence. Our group are developing a gene therapy method for Duchenne Muscular Dystrophy which involves the introduction of a dystrophin cDNA, by direct intramuscular injection. The problem thus far, is that only a small percentage of the fibres are transfected and would be thus rescued in this way. The possibility also exists of using the muscle as an *in vivo* expression system and introducing the coding sequence of a hormone, growth factor or clotting factor providing a signal sequence is included in the engineered gene. This would ensure that the protein product is exported from the muscle cells to elevate the serum level.

Muscle stem cells offer the possibility of not only being used to introduce new genes into muscle fibres but they may in the future be used to "re-seed" and rebuild tissue. Skeletal muscle is a post-mitotic tissue and once fusion of the myoblast has taken place during embryonic differentiation no further mitoses takes place. Muscle fibres increase in length by about three times and the extra nuclei are donated by satellite cells. The origin of these is not known but they may be a type of residual stem cell. Cardiac muscle does not have satellite cells but it is possible to transform totipotent cells into cardiac myoblasts and thus one can visualize sometime in the future rebuilding the myocardial by injecting totipotent cells with appropriate growth factors into the pericardium to make new heart tissue. The possibility therefore exists of using these muscle stem cells not only to reconstruct muscle tissue but also for gene therapy as it is a way of introducing new genes.

PRESENT AND FUTURE APPLICATIONS

1. *Surgical limb lengthening*

This provides a very basic but very effective means of engineering skeletal muscle. Bone distraction is used in orthopaedic centres to correct major deformities and limb length discrepancy. The procedure involves osteotomy followed by distraction of the two parts of the bone using an external frame. New bone forms between the cut ends and the muscle also grows in length by producing more sarcomeres in series. The remodelling of the collagen framework of the muscle was somewhat slower and in older children this seems to be the rate limiting factor for the addition of new sarcomeres. If the distraction of the two bone shafts is too rapid the connective tissue and the myofibrils are not pulled out evenly and some sarcomeres become too long and thus the muscle is weakened. (Simpson *et al.*, 1991) Nevertheless, the results indicate good adaption to the imposed length providing the distraction rate is not too rapid (Williams *et al.*, 1991). However, we have yet to fully investigate the effectiveness of a distractor driven by a linear motor that provides continuous stretch rather than the manual versions that provide stepwise limb length increments.

2. *Neosphincter*

N S William's Group at the Royal London Hospital surgical Unit have pioneered the construction of anal sphincters using the gracilis muscle. The muscle is stimulated with a specially designed pacemaker to remain contracted and generate enough force to close the anal canal. The pacemaker can be switched off during defecation using a small magnetic switch, the problem is the length of time it takes the gracilis muscle to adapt to its new role by becoming fatigue resistant (Williams *et al.*,

1991). Even though the conditions have not been fully optimised, the gracilis muscle does transform within a period of 6 months to a year, from a predominantly fast muscle to one that can maintain force continuously for long periods of time. This shows the extreme adaptability of skeletal muscle.

3. *Cardiac assist*

Surgical procedures for "cardiac assist" myoplasty have been developed by several teams in the States and Europe Bridges *et al.,* 1989. The latissimus dorsi muscle is used as a contractile wrap around the ventricles (Carpentier *et al.,* 1991, Millner 1991) the aorta (Pattison *et al.,* 1991) or used to fashion a separate ventricle (Bridges *et al.,* 1989). The muscle flap is translocated with its nerve and blood supply intact and is activated by a specially designed pacemaker which is triggered by the QRS complex. The muscle fibres adapt quite rapidly adapt to their changed activity pattern and become considerably more fatigue resistant (Cumming *et al.,* 1991). The main problem is that when the LD flat is translocated to form a ventricle or aortic wrap there is a marked decline in contractile power output. This is attributable to muscle fibre atrophy and also the slowing of the contractile apparatus. As mentioned, the stretch effect and the adaptation to an increased length is known to be associated with increased protein synthesis (Goldspink D.F. *et al.,* 1991). Recently we have studied the way gene expression of skeletal muscle is influenced by stretch and/or electrical stimulation. Briefly, repetitive contractions cause the switch in expression from fast genes to slow genes and when the stimulation is combined with the stretch then the muscle fibres become completely reprogrammed (Goldspink et 1992). The prospects of engineering the tissue so that it has the appropriate characteristics so that it can be used for surgical construction of autologous devices to provide force and power as in neosphincters and cardiac assists, looks promising because muscle is so adaptable and malleable.

4. *Muscle as an in vivo expression system for gene transfer*

Recently, Jon Wolff's group in the States published a very interesting finding in that they found that following a single injection of certain gene constructs into skeletal muscle, the DNA was taken up and expressed by a percentage of muscle fibres (Wolff *et al.,* 1990). This has now been confirmed for skeletal muscle and cardiac muscle of several species, but for no other cell types. Our group were initially sceptical but we have found that the method works well for fish (Hansen *et al.*, 1991) as well as for mammalian skeletal muscle (Wells and Goldspink 1992). Indeed, they obtained much higher levels of expression than Wolff's group as they used gene constructs that were under the control of muscle specific promoters. It was also found that the injected constructs gave much higher expression in younger muscles (Wells and Goldspink 1992). The implications of this are that gene therapy using this approach should be carried out at as young an age as possible. The genes introduced in this way remain in plasmid form and do not integrate with the chromosomes which means that there is no risk of disrupting clusters of developmental genes in the host's genome with the risk of cancer formation etc. It is for this reason that the emphasis in gene therapy development is to construct and introduce mini-chromosomes. The simplest kind of mini-chromosome is a plasmid construct and rather fortuitously these can be introduced into skeletal muscle very simply by intramuscular injection. The expression of the plasmid constructs is of long duration so repeated injections should not be necessary. Measurements over several months following injection and no fall off in expression has been detected. The different conditions and hence the variable levels of expression have made definitive measurements difficult. Also what determines the expression life of the introduced DNA in plasmid form still needs to be elucidated.

One problem at the present time is that only a small percentage of the muscle fibres are transfected following a single intramuscular injection and this is very variable. A low percentage transfection efficiency is presumably adequate when one is using the muscle to produce low constituent levels of a clotting factor, a growth factor or a hormone. However, if one is attempting to salvage muscle fibres eg. in muscular dystrophy, then as high a percentage of muscle fibres as possible should be transfected following a single injection. New methods such as using liposomes promise to improve the efficiency of the gene transfer. Perhaps more urgently, we need to know how the physical status of the muscle tissue affects the level of uptake and expression of the different designs of gene construct. These findings should have considerable implications for somatic cell gene transfer for a number of diseases as it should be possible to have any coding sequence driven by a muscle type promoter. For example, the coding sequence for any hormone or growth factor might be spliced to a muscle promoter and other appropriate sequences such as a leader (signal) peptide sequence. In this way the serum levels of that hormone could be elevated. Also recombinant DNA vaccines might be introduced providing again an appropriate signal sequence is included to ensure that the antigen is exported from the muscle cells. Direct muscle gene injection therefore offers the prospects of carrying out gene therapy in a minimally invasive and controllable fashion.

5. *Muscle stem cells and myoblast transfer*

Myoblast transfer was first described by Partridge *et al.*, (1978) and Watt *et al.*, (1981) who implanted enzyme disaggregated muscle precursor cells into regenerating mouse muscle. They found that the mononuclear myoblasts fused with the hosts developing fibres despite being derived from a genetically different inbred strain of mice. This offered the possibility of using myoblasts to transfer genes into muscle fibres in such conditions as Duchenne muscular dystrophy. This disease results from a defect in a gene that has been called the dystrophin gene. Dystrophin protein is required in relatively low amounts to apparently stabilize the muscle fibre membrane. The fact that the muscle fibre is a synctium means that the gene products from the genetically normal nuclei should be transported and assembled all along the muscle fibre. Thus a few myoblasts with the normal dystrophin gene fusing with each muscle fibre should be sufficient to produce enough dystrophin to protect that fibre from physical damage. This technique has therefore been termed myoblast transfer therapy (Partridge 1988) and trials have been carried out on human patients (Law *et al.*, 1990, Karpati *et al.*, 1990, Gussoni *et al.*, 1992). The problem at the present time is that the cells used are myoblasts of a cell line such as C2/C12 and hence there is always the fear that the transformed cells may become malignant in the patient. True muscle satellite cells are difficult to handle as they must be maintained in primary culture and they do not transfect easily. Unless the patients own cells are used there will be the problem of the immune response and powerful immuno-suppressing drugs and procedures have to be employed. Therefore the strategy is to develop a muscle stem cell that will not produce tumours or not evoke an immune response. Within the next decade or so it is possible that cardiac and skeletal muscle stem cell lines will be developed so that these can be used to reconstruct tissue and introduce normal genes and in the case of agricultural animals to introduce growth factor genes that result in genetic enhancement and improvements in meat quality and quantity.

ACKNOWLEDGEMENTS
This work was supported by grants from the Wellcome Trsut, The AFRC and Action Research of the UK.

REFERENCES

Cumming D.V.E., Pattinson C.W., Lovegrove C.A., Dewar A., Dunn M.J., Yacoub M. and Goldspink G. (1991): Biochemical and structural adaption of autologous skeletal muscle used for counter pulsation. Int. Journal of Cardiology, **30**: 181-190.

Fischman D.A. (1986): Myofibrillo-genesis and morphogenesis of skeltal muscle. In: *Myology, Basic and Clinical.* Eds A.G. Engel and B.Q. Banker. McGraw Hill Co. New York pp 5-39.

Goldspink G. (1968): Sarcomere length during the post-natal growth of mammalian muscle fibre. J. Cell. Sci. **3**: 539-548.

Goldspink G. (1971): Changes in striated muscle fibres during contraction and growth with particular reference to the mechanism of myofibril splitting. J. Cell. Sci., **9**: 123-138.

Goldspink, G. (1977): Muscle energetics and animal locomotion. In: *Mechanics and Energetics of Animal Locomotion,* edited by R.M. Alexander and G. Goldspink. London: Chapman and Hall, p. 57-81.

Goldspink G. (1985): Malleability of the motor system: a comparative approach. J. Exp. Biol. **115**: 375-391.

Goldspink G. (1984): Alterations in myofibril size and structure during growth, exercise, and changes in environmental temperature. In: *Handbook of Physiology. Skeletal Muscle*, Editors: Peachy L.D., Adrian R.H. and Geiger S.R., Bethesda, MD: Am. Physiol. Soc., 1983, sect. 10, chapt. 18: p. 539-554.

Goldspink D.F., Easton F., Winterburn J., Williams P.E. and Goldspink G. (1991): The role of passive stretch and repetitive electrical stimulation in preventing skeletal muscle atrophy while reprogramming gene expression to improve fatigue resistance. Journal of Cardiac Surgery, **6** 218-226.

Goldspink G., Scutt A., Loughna P., Wells D., Jaenicke T. and Gerlach G-F. (1992): Gene expression in skeletal muscle in response to mechanical signals. Am. J. Physiol. **262** R356-R363.

Gussoni E., Pavlath G.K., Lanctot A.M., Sharma K.R., Miller R.G., Steinman L. and Blau H.M. (1992): Normal dystrophin transcriptsdetected in Duchenne Muscular dystrophy patients after myoblast transportation. Nature, **356**: 435-348.

Hansen E., Goldspink G., Butterworth P.W. and Chang K-C. (1991): Strong expression of some mammalian gene constructs in fish muscle following direct gene transfer. FEBS Letts, **290**: 73-76.

Karpati G. and Poultiot Y.C. Holland Carpenter Sand Holland. (1990): Implantation of non dystrophy allogenic myoblasts into dystrophic muscle of mdx mice produces mosaic fibres of normal microscopic phenotype. Cellular and Molecular Biology of Muscle Development. pp 973-985.

Law P.K., Bertorini T.E., Goodwin T.G., Chen M., Fang Q., Li H.J., Kirby D.S., Florendo J.A., Herrod H.G. and Golden G.S. (1990): Dystrophin production induced by myoblast transfer therapy in DMD. The Lancet, **336**: 114-115.

Loughna P.T., Goldspink D.F. and Goldspink G. (1986): Effect of inactivity and passive stretch on protein turnover in phasic and postural rat muscles. J. Appl. Physiol., **61**: 173-179.

Loughna P.T., Izumo S., Goldspink G. and Nadal-Ginard B. (1990): Disuse and passive stretch cause rapid alteration in expression of development and adult contractile protein genes in adult skeletal muscle. Development, **109**: 217-223.

McMinn R.H.M. and Vrbova G. (1964): The effect of tenotomy on the structure of fast and slow muscle in the rabbit. Quant. J. Exp. Physiol. **49**: 424-430.

Partridge T.A., Grounds M. and Sloper J.C. (1978): Evidence of fusion between host and donor myoblasts in skeletal muscle grafts. Nature, **273**: 306-308.

Partridge T.A., Morgan J.E., Coulton G.R., Hoffman E.P. and Kunkel L.M. (1989): Conversion of mdx myofibres from dystrophin-negative to -positive by injection of normal myoblasts. Nature, **337**: 176-179.

Pattison C.W., Cummings D.V.E., Clayton-Jones D.G., Goldspink G., Dunn M.J. and Yacoub M.H. (1989): Variable adaptation of molecular mechanisms in relation to the use of autologous striated muscle to augment myocardial function. Cardiovasc. Res., **23**: 593-600.

Simpson, H., Kyberd, P., Goldspink, G. and Kenwright, J. (1991): Muscle development during limb lengthening. Report of the Oxford Orthopaedic Centre. pp. 25-27.

Tabary J.C., Tabary C., Tardieu C., Tardieu G. and Goldspink G. (1972): Physiological and structural changes in the cat's soleus muscle due to immobilization at different lengths by plaster casts. J. Physiol. **224**.

Watt D.J., Partridge T.A. and Sloper J.C. (1981): Cylcosporin A is a means of providing rejection of skeletal allographs in mice. Transplantation **31**: 266

Wells D.J., Wells K.E., Walsh F.S., Goldspink G., Love D.R., Chan-Thomas P, Dunckley M., Piper T. and Dickson G. (1992): Human dystrophin expression corrects the myopathic phenotype in transgenic mdx mice. Human Molecular Genetics. **1** 35-40

Wenderoth M.P. and Eisenberg B.R. (1987): Incorporation of nascent myosin heavy chains into thick filaments of cardiac myocytes in

Williams P. Griffin G.E. and Goldspink G. (1971): Region of longitudinal growth in striated muscle fibres. Nature New Biology, **232**: 28-29. thyroid treated rabbits. J Cell Biol. **108**: 2771-2780.

Williams R.S., Salmons S., Newsholme, A., Kaufman R.E. and Mellor J. (1986): Regulation of nuclear and mitochondrial gene expression by contractile activity in skeletal muscle. J. Biol. Chem. **261**: 376-380.

Williams N.S., Patel J., George B.D., Hallan R.I. and Watkins E.S. (1991): Development of an electrically stimulated neonatal sphincter. Lancet, **338**: 1166-1169.

Williams P.E., Simpson, A.H.R.W., Kyberd, P., Kenwright, J. and Goldspink, G. (1992): Physiological and structural changes in the rabbit anterior tibialis muscle following surgical limb-lengthening. J. Physiol. **446**, 571P.

Wolff J.A., Malone R.W., Williams P., Chong W., Acsadi G., Jani A. and Felgner P.F. (1990): Direct gene transfer into mouse muscle *in vivo*. Science Wash. DC. **247**: 1465-1468.

SMALL INTESTINAL SUBMUCOSA (SIS): A BIOMATERIAL CONDUCIVE TO SMART TISSUE REMODELING

Stephen F. Badylak

From the Hillenbrand Biomedical Engineering Center, Purdue University, W. Lafayette, IN 47907,

Address for reprints: Dr. Stephen F. Badylak,
Hillenbrand Biomedical Engineering Center,
1293 Potter Engineering Center
Purdue University,
W. Lafayette, IN 47907-1293

INTRODUCTION

Tissue Engineering has been defined as "the application of the principles and methods of engineering and life sciences toward the fundamental understanding of structure-function relationships in normal and pathological mammalian tissues, and the development of biological substitutes to restore tissues" (Skalak and Fox, 1988). This discipline involves structure-function relationships at the molecular, intracellular, intercellular, tissue and whole organ levels and attempts to bridge the gap between studies at the cell level and those at the tissue-organ level (Heineken and Skalak, 1991). The studies described below were targeted at the whole tissue response and somewhat less at the individual cell level. This manuscript describes the findings from a series of in-vivo studies in which a novel biomaterial derived from the small intestinal submucosa contributed to the remodeling of a variety of tissues. A definitive explanation for the results of these studies is not possible with our current understanding of the regulation of tissue remodeling; however, speculation for some of the structure-function observations is presented. This material was presented at the Tissue Engineering Symposium in Keystone, Colorado in April, 1992.

BACKGROUND INFORMATION FOR A NEW BIOMATERIAL

Work has been conducted at the Hillenbrand Biomedical Engineering Center during the past 6 years with an obscure biomaterial taken from the small intestine of various mammalian species. This material is referred to as small intestinal submucosa (SIS). A more detailed description of its structure is presented below. During the course of work, a considerable body of knowledge has been learned about the response of the body to implanted SIS, the morphology and chemical composition of the native material, and the potential for adult differentiated tissues to remodel given the appropriate stimuli.

SIS consists of a trilaminar portion of the small intestine including the stratum compactum layer of the tunica mucosa, the tunica muscularis mucosa (not present in all species), and the tunica submucosa. These layers are harvested in a manner which has been previously described (Badylak et al, 1989; Sandusky et al, 1992). This thin, approximately 100 μM thick, material has several interesting properties. The stratum compactum surface is surprisingly thromboresistant when exposed to flowing blood whereas the opposite surface is thrombogenic (Lantz et al, 1990). The material has a porosity index of 0.2 ml/cm^2 min^{-1} when tested from the stratum compactum surface toward the submucosal surface but an 4-fold greater porosity index in the opposite direction (Hiles et al, in press; Ferrand et al, in press). SIS has viscoelastic properties and tensile properties that provide for greater hoop directional strength than longitudinal directional strength. It has been determined in our laboratory that SIS is 87% water and that approximately 40% of the dry weight is collagen (by hydroxyproline assay).

Microscopic examination of SIS shows that the stratum compactum layer consists of more densely organized connective tissue than the underlying submucosal layer but only 10% to 15% of the thickness of the material is occupied by the stratum compactum. SIS is relatively acellular and the bulk of the material consists of extracellular connective tissue matrix. The intact cells present in SIS consist of occasional fibrocytes and the endothelial cells which line the vascular channels which once coursed through these layers of the intestine.

The response of mammalian tissues to SIS provides a fascinating finding for those interested in tissue engineering and a glimpse of the potential for tissue and organ replacement. A chronologic overview of selected studies conducted in our laboratory is presented below. This information should contribute to the body of knowledge available to tissue engineers.

ARTERIAL GRAFT STUDIES

The earliest study of SIS involved its use as a large diameter (approximately 8.0 mm ID) autograft in which a 5.0 cm long segment of infrarenal aorta was replaced in the dog (Badylak et al, 1989). Thirteen dogs were implanted and two died as a result of technical errors within the first 2 weeks of the study. Ten of the remaining 11 dogs were sacrificed at various times and the grafted material was evaluated. All grafts remained patent and developed a smooth endothelial lining and a 500 to 700 μM thick wall. There was no evidence for aneurysmal dilatation, intimal hyperplasia, infection or mineralization at any time. One dogs remains alive more than 5 years after surgery and is free of any of the above mentioned complications. The significance of this early study was the apparent nonthrombogenic nature of a collagen surface, the remodeling of the native implant into a new vascular channel, and the ability of SIS to support endothelial cell growth.

Following the initial success of SIS as a large diameter autograft, it was tested as a small diameter (4.3 mm ID) arterial graft in the dog (Lantz et al, 1990). Eighteen dogs had bilateral femoral or carotid artery SIS grafts implanted (total of 36 grafts). Two grafts ruptured within 3 days of surgery and 5 grafts occluded within the first 2 weeks. The overall patency rate was therefore 80%. Similar to the large diameter graft study, there was no evidence for infection, intimal hyperplasia or mineralization. Three dogs remain alive with patent grafts after 4 years. During the course of this study, it was observed that the connective tissue which developed in the wall of the graft organized in a primarily circumferential direction and became richly vascularized early in the course of remodeling (i.e., 4 to 7 days). The initial host response to the implanted SIS was characterized by the classic inflammatory reaction that would be expected for an injured connective tissue. More specifically, there was initial tissue edema, neutrophil accumulation which changed to mononuclear cell predominance by 48 hours, followed by the presence of spindle cells and a deposited eosinophilic staining connective tissue matrix. The thickness of the remodeled graft stabilized at approximately 600 μM and subsequent studies have repeatedly shown this final tissue architecture to be identical for all arterial grafts. However, unlike the typical response to injury for a specialized tissue in which scar tissue might be expected to form, a continued remodeling of the connective tissue occurred and more specialized organization ensued which is described in the following study.

Additional small diameter graft studies were performed with allogenous and xenogeneic (porcine) SIS tissue because of the practical limitations of an autogenous SIS graft for any eventual clinical application (Sandusky et al, 1992). Twenty-four SIS grafts (porcine origin) were placed in the carotid artery location of dogs and autogenous saphenous vein was used as a graft for the contralateral carotid artery in each dog. Twenty of 24 SIS grafts and 21 of the 24 saphenous vein grafts remained patent throughout this 6 month study. Serial evaluation of the grafts was performed and several significant observations were made which distinguished this study from the previous small diameter graft study. First, xenogeneic SIS tissue elicited the identical response as the autogenous tissue which was used in the original study and showed no evidence of an adverse immunologic response. Second, by 14 days after surgery, endothelial cells covered the luminal surface and were staining positive for vWF. Third, like the saphenous vein grafts at 90 days, the xenogeneic SIS grafts had arterialized with an intima covered by endothelium, a smooth muscle media (positive staining with HHF-35 antibody against actin), and marked adventitial fibrosis. Stated differently, this study was the first to characterize the remodeling of SIS with respect to cell types and tissue organization that were appropriate for the location in which it was placed. Xenogeneic SIS has since been placed in the femoral artery location of 4 rhesus monkeys (approximately 2.0 mm ID). One monkey has been sacrificed to date and examination of the graft shows remodeling with smooth muscle cells, intima, and endothelial cells like that observed in the dog study.

VENOUS GRAFT STUDY

To determine the remodeling response of SIS in a vascular location which is devoid of high blood pressure or pulsatile flow, we replaced the superior vena cava in 9 dogs with SIS (Lantz et al, in press). One dog exsanguinated within 24 hours of surgery and 6 of the remaining 8 dogs were sacrificed up to 17 months after surgery. The explanted grafts had remodeled into thin walled (approximately 100 μM thick), endothelialized vascular channels which appeared morphologically identical to a normal vein. There were no actin containing spindle cells nor was there evidence for the original SIS material in the new vessel. Two dogs remain alive after more than 3 years. This study is significant with respect to tissue engineering because the remodeled blood vessel appeared and functioned in a fashion appropriate for the location in which it was placed (i.e., venous location). Stated differently, the tissue environment and stressors upon the SIS were different from the previous arterial implants and the response of the host to the implant remodeling reflected these differences.

INTERIM CONCLUSIONS AND OBSERVATIONS

Collectively, the above studies showed that SIS can be remodeled in-vivo into blood vessel substitutes that are more than simply organized scar tissue. It seems clear that at least one xenogeneic tissue can: 1) survive in a recipient host without adverse immunologic consequences; 2) elicit connective tissue remodeling that is responsive to local body stressors and microenvironment; 3) support the controlled proliferation of relatively specialized tissues such as smooth muscle and endothelial cells; and 4) induce a bodily response including neovascularization that is resistant to infection.

INFECTION STUDY

The SIS material used in the vascular graft studies was nonsterile and yet clinical or laboratory signs of infection or morphologic evidence for tissue infection were not found in a single animal. A study was conducted which evaluated the response of SIS and host tissues to deliberate massive infection (Badylak et al, in press). Eighteen dogs were divided into 2 equal groups and the infrarenal aorta was replaced with either polytetrafluoroethylene (PTFE) or SIS graft material. One hundred million *S. aureus* organisms were deposited directly on the graft at the time of surgery and the dogs observed for 30 days. One dog with a PTFE graft died of hemorrhage from an anastomosis site at 21 days. Of the remaining 8 dogs with PTFE grafts, 4 had positive cultures results from the explanted graft material and all had histologic evidence for persistent infection. These dogs also had chronic fever and increased white blood cell counts at day 30. All 9 dogs with SIS grafts had patent grafts, were afebrile after the first week, had normal white blood cell counts at day 30, had negative culture results, and had the histologic appearance of graft remodeling with collagen that was free of active inflammation. It was concluded that large diameter arterial SIS grafts were more resistant to persistent infection with *S. aureus* than PTFE grafts in this dog model of deliberate bacterial inoculation. We speculate that the rich and rapid vascular response that occurs in dogs implanted with SIS is important in the infection resistance that has been observed in all studies with SIS.

SOFT TISSUE ORTHOPEDIC STUDIES TO DETERMINE THE FATE OF IMPLANTED SIS

During the past 2 years it has been shown that the remodeling properties and infection resistance properties of SIS extend beyond those observed in the vascular system (Aiken et al, in press). Various configurations of the material have been used to replace the anterior cruciate ligament, the Achille's tendon, and the medial collateral ligament in xenograft dog studies. The term "configuration" refers to: 1) the orientation of the material with regard to stratum compactum surface vs. the submucosal surface; and, 2) the use of the material in its natural tube shape vs. sheets of the material vs. 1.0 cm wide strips of the material braided into ropes.

Twenty-five dogs had one of their anterior cruciate ligaments (ACL) completely excised then replaced with porcine origin SIS. Eleven dogs had the SIS configured as a simple tube with the stratum compactum surface facing inward as it was implanted; ten dogs had sheets of SIS rolled into 4.0 cm wide solid tubes for implantation; and 4 dogs had braided strips of material implanted in the ACL location. Although the initial longitudinal strength of the various configurations differed greatly dependent upon the amount of material present and its arrangement into tubes vs. braids, by 4 months after surgery the remodeled ligaments all had very similar tensile strength properties and morphology. Significant with regard to tissue engineering, the whole organism response to the SIS template was apparently such that signals from the surrounding tissue milieu ignored the implanted configuration of the SIS and responded to the needs of the body for connective tissue in this particular intraarticular location.

The tissue response to the implanted material showed a rapid neovascularization, deposition of early disorganized collagenous connective tissue, followed by gradual organized collagen along the lines of tensile load. The remodeled connective tissue supported the exterior growth of synovial cells on the surface of the neoligament. The portion of the SIS which passed through bone tunnels to attachment sites showed replacement of the SIS with new bone within the tunnels and fibrocartilage at the bone-ligament interface. Once again, the body deposited the appropriate type of connective tissue in the location most suited for that type of connective tissue. This might be referred to as "smart remodeling".

Since the mass of material present in the remodeled ligament was markedly greater than that of the originally implanted SIS, it is obvious that the original SIS material represented only a small part, if any, of the neoligament. The same conclusion can be made about the previously described blood vessel studies in which new vascular channels were formed. In an effort to determine the fate of implanted xenogeneic SIS in a recipient host, a monoclonal antibody (MoAB) specific for porcine SIS was developed. The relative nonimmunogenicity of the material was confirmed by the difficulty encountered in isolating such an antibody.

With the use of the MoAB to porcine SIS, it was possible to determine the length of time that the original material was still present in the remodeled ligaments. By examining tissue sections over time, it was clear that there was gradual deterioration of the implanted native SIS beginning immediately after implantation and the ingrowth of host tissues from all directions. By 60 days, most ligaments contained only remnants of the original implant and by 90 days, none of the remodeled ligament stained positively with the MoAB.

A similar experiment to the ACL replacement was conducted in which 10 dogs have had one of their Achille's tendons replaced with porcine SIS. A 2.0 cm long section of the tendon originating just distal to the musculotendinous junction was excised and the gap replaced by a single tube of SIS. A partial weight bearing external brace was used for 5 weeks followed by complete weight bearing in a spacious animal holding facility. All 10 of the dogs showed a rapid replacement of the SIS with a solid tube of connective tissue that stabilized at a diameter equal to that of the adjacent normal tendon.

Histologic examination showed the imperceptible blending of normal longitudinally aligned collagen with the neoligament by 6 to 8 weeks after surgery. There was a complete absence of native SIS as determined by MoAB staining by 8 weeks. With the exception of small granulomas around the suture placement sites, there was complete absence of any inflammatory response. The original SIS material, identified with the MoAB, was surrounded by a rich vascular supply which extended to the surface of the ligament. There was a normal synovial lining and tenosynovium by 6 weeks after surgery. The tensile strength of the remodeled ligament was at 70% of the contralateral normal ligament by 8 weeks after surgery and was at 100% by 12 weeks after surgery. In 2 dogs in which the brace was kept in place for 8 weeks instead of 5 weeks, the remodeling process was slowed as evidenced by a weaker tendon. These 2 dogs had remodeled tendons which had 40% and 45% of the contralateral tendon strength at 8 weeks.

This study helped to confirm many of the properties of SIS observed in the other studies. Neovascularization, support of epithelial cell growth on appropriate surfaces, absence of immune rejection, and extensive connective tissue matrix deposition in and surrounding the site of the SIS implant are all features common to the host response to SIS. This study also demonstrated the ability of SIS to develop strong connective tissue in a short time frame when the stress of the animal's weight bearing was applied.

IN-VITRO CELL CULTURE STUDIES WITH SIS

Tissue Engineering at the tissue and organ level as described in the above studies is logically the result of earlier events at the intracellular, cellular, and intercellular levels. Although studies are in progress to determine the effect of SIS upon cell to cell communication and cytokine interactions, no definitive data is available at this time. It is known however, that sterilized SIS will support the in-vitro cell growth of many cell types including: keratinocytes, endothelial cells, human epidermoid carcinoma cells, Chinese hamster ovary cells, rabbit aorta smooth muscle cells, human bone cells, and colon carcinoma cells.

SUMMARY

SIS provides an opportunity for the in-vivo and in-vitro study of tissue engineering principles. It appears that relatively specialized connective tissue such as smooth muscle can be stimulated to proliferate in an *organized* fashion in at least some adult mammals (Stone et al, 1992; Yannas and Burke, 1980; Yannas, 1988). Furthermore, the inducers of this process may not need to persist after connective tissue remodeling is in progress. The regulators and controllers of tissue remodeling appear to be active when SIS is implanted and this material may provide a model system for the study of such regulators and controllers.

REFERENCES

Aiken SW, Badylak SF, Toombs JP, Shelbourne KD, Hiles MC, Lantz GC, Van Sickle D. Small intestinal submucosa as an intra-articular graft material in dogs. *J Surg Res* (in press).

Badylak SF, Lantz G, Coffey A, Geddes LA. Small intestinal submucosa as a large diameter vascular graft in the dog. *J Surg Res* 1989; 47:74-80.

Badylak SF, Coffey AC, Lantz GC, Tacker WA, Geddes LA. Infection resistance of small intestinal submucosa (SIS) vs. polytetrafluoroethylene vascular grafts in a dog model. *J Vasc Surg* (in press).

Ferrand BK, Kokini K, Badylak SF, Geddes LA, Hiles MC, Morff RJ. Directional porosity of porcine small-intestinal submucosa. *J Biomed Materials Res.* (in press).

Heineken F, Skalak R. Tissue engineering: a brief overview. *J Biomech Eng* 1991; 113:111.

Hiles MC, Badylak SF, Geddes LA, Kokini K, Morff RJ. Porosity of porcine small intestinal submucosa for use as a vascular graft. *J Biomed Materials Res* (in press).

Lantz G, Badylak SF, Coffey A, et al. Small intestinal submucosa as a small diameter arterial autograft in the dog. *J Invest Surg* 1990; 3:217-227.

Lantz GC, Badylak SF, Coffey AC, Geddes LA, Sandusky GE. Small intestinal submucosa as a superior vena cava graft in the dog. *J Surg Res* (in press).

Sandusky GE, Badylak SF, Morff RJ, Johnson WD, Lantz G. Histologic findings after in-vivo placement of small intestine submucosal vascular grafts and saphenous vein grafts in the carotid artery in dogs. *Am J Pathol* 1992; 140:317-324.

Stone KR, Rodkey WG, Webber R, McKinney L, Streadman R. Meniscal regeneration with copolymeric collagen scaffolds. *Am J Sports Med* 1992; 20:104-111.

Yannas IV, Burke JF. Design of an artificial skin. I. Basic design principles. *J Biomed Mater Res* 1980; 14:65-81.

Yannas IV. Regeneration of skin and nerve by use of collagen templates. In: Nimni ME (ed); Collagen Biotechnology, Vol III; Ch 30. Boca Raton, FL, CRC Press, 1988.

MATRIX ENGINEERING: REMODELING OF DENSE FIBRILLAR COLLAGEN VASCULAR GRAFTS IN VIVO.

Crispin B. Weinberg, Kimberlie D. O'Neil, Robert M. Carr, John F. Cavallaro, Bruce A. Ekstein, Paul D. Kemp, Mireille Rosenberg
Organogenesis Inc.

Jose P. Garcia, Michael Tantillo, Shukri F. Khuri
West Roxbury Veterans Administration Medical Center

INTRODUCTION

Autogenous vessels, primarily saphenous veins, remain the "gold standard" for arterial grafting (Kent et al, 1989); however, they do not function as normal blood vessels immediately after surgery. In both animals (Jones et al, 1973; Fonkalsrud et al, 1978) and humans (Unni et al, 1974; Spray & Roberts, 1977) there is extensive endothelial and smooth muscle cell necrosis, a platelet-fibrin deposit on the intimal surface, and medial edema shortly after grafting. A vein graft functions as a scaffold for extensive remodeling resulting in an artery-like structure with an endothelium and smooth muscle cell media after a period of weeks to months. Thus cardiovascular surgeons have, perhaps unknowingly, been successfully practicing tissue engineering for several decades. Similarly, other biological materials, such as the submucosa of the small intestine, can also serve as scaffolds for remodeling and arterialization (Badylak *et al*, 1989; Sandusky *et al*, 1992).

We have used Type I collagen to fabricate, *in vitro,* a construct designed to function as a vascular graft that will guide remodeling in an orderly manner, *in vivo.* Until recently it has not been possible to fabricate *in vitro* collagen constructs with long native-banded fibrils (Bruns and Gross, 1985; Carr *et al*, 1991; Weinberg *et al*, 1991). Thus earlier constructs have not had sufficient strength for mechanically demanding applications such as vascular grafts (Gross and Kirk, 1958; Chvapil *et al*, 1973; Weinberg and Bell, 1986). By concentrating collagen around a mandrel prior to fibril formation, we produce a collagen matrix of long, but fine, native-banded collagen fibrils at densities approaching those of normal tissues. By controlling the Ph at which this dense fibrillar collagen (DFC) matrix is laid down, we can make a smooth, relatively non-thrombogenic surface as shown by acute platelet uptake studies and short term vascular implants (Carr *et al*, 1991; Weinberg *et al*, 1991).

In this paper we present implantation experiments showing that the DFC matrix can be remodeled *in vivo* to produce tissue-like structures in the course of a few months. In order to provide sufficient strength for suturing and to maintain the structure during the remodeling process, the grafts were made with an integral knit mesh. Most experiments utilized a Dacron™ polyester mesh, but recently we have adapted the DFC process to extrude collagen threads (Cavallaro *et al*, 1991) and have knit them into a mesh which can be crosslinked to maintain strength and then integrated with a DFC matrix.

Porosity has long been known to be crucial for vascular graft healing (Wesolowski *et al*, 1961) and has recently been shown that endothelialization can be promoted by pores that allow transmural capillary ingrowth (Clowes *et al*, 1986; Golden *et al*, 1990). We can provide the DFC matrix with pores by laser drilling. These pores facilitate ingrowth of endothelial cells in tissue culture and preliminary implant studies suggest that they result in endothelialized transmural channels.

MATERIALS AND METHODS

Dense Fibrillar Collagen

Type I bovine collagen was prepared and formed into DFC sleeves as described previously (Carr *et al*, 1991). Briefly, collagen was concentrated *in situ* around a cylindrical microporous ceramic mandrel by perfusing the mandrel's lumen with 20% (weight/volume) polyethylene glycol (PEG). PEG was dissolved in 0.05% acetic acid (Ph 3.8) or 50mM phosphate buffer (Ph 7.1). Both the acidic and neutral concentrating solutions were adjusted to 500-600 mOsM with NaCl. The resulting tube of highly concentrated collagen was air-dried and then rehydrated in phosphate-buffered saline (PBS).

Meshed sleeves were made by interrupting the collagen deposition process, slipping an open (1-2mm holes) knitted mesh over the partially formed sleeve, and then continuing the collagen deposition. In most experiments a polyester (Dacron™) mesh was used while more recently we employed a mesh of collagen threads prepared and crosslinked as described elsewhere (Cavallaro *et al*, 1991).

In a preliminary study to examine the effect of DFC porosity on remodeling, air-dried grafts were laser drilled using a pulsed UV excimer laser. A stainless steel mask was employed to produce a regular pattern of holes 15-30μm in diameter with a center to center spacing of about 500μm. The grafts were rehydrated in PBS before use.

Surgical Procedures

Mongrel dogs (20-25kg) were anesthetized with 1.0% (vol/vol) halothane. DFC vascular grafts (3.5-4.0mm I.D. x 5cm) were implanted as interposition grafts in the left superficial femoral artery using 6-0 polypropylene sutures. The graft was kept moist throughout the procedure and handled as gently as possible. Heparin (200 U/kg) was administered prior to clamping the femoral artery and was reversed with protamine sulfate, if needed. One aspirin (325mg) was given daily post-operatively and an antibiotic (Kefzol, 1gm) was administered peri-operatively.

Grafts were monitored by positive contrast angiography as well as by Doppler ultrasound on a regular basis. Explanted grafts were cut into 5 segments (1-2 cm long), photographed, and fixed in buffered formalin.

Animal care complied with the "Principles of Laboratory Animal Care" and the "Guide for Care and Use of Laboratory Animals" (NIH Publication No. 80-23, Revised 1985).

Histology & Morphology

Segments of explanted graft were embedded in paraffin. Transverse sections were cut every 500 μm through each anastomosis and every cm through the center of the graft, stained with hematoxylin and eosin, and scored on a standardized histopathology form. Staining with *Ulex europaeus* lectin was used to identify endothelial cells. Staining with antibody to smooth muscle actin was to identify cells with a smooth muscle phenotype (smooth muscle cells, myofibroblasts, or pericytes). Scanning electron microscopy of DFC matrices was performed by conventional methods.

RESULTS

Meshed dense fibrillar collagen grafts are tubes that have burst strengths above 500mm Hg, have low water permeability, and are moderately compliant. We have previously shown that DFC sleeves made at neutral Ph have a smooth surface and relatively low thrombogenicity as evaluated by platelet uptake studies and short term (1 week) arterial implants. In contrast, acidic sleeves have a fibrillar surface and are quite thrombogenic (Carr *et al*, 1991; Weinberg *et al*, 1991).

While many DFC grafts in these initial studies failed due to thrombosis or hemorrhage (usually 6-20 days after grafting), eleven grafts remained patent until scheduled explant at 1 month or 3 months. Graft diameters at time of explant, determined from arteriograms, were 4.0 $\pm$ 0.4mm (mean $\pm$ S.D). Due to an improved knit configuration, these grafts maintained their diameter much better than those in an earlier study of acidic DFC implants in sheep.

At one month regions where the DFC on the luminal surface remained intact (up to 75% of the graft surface) had little or no platelet-fibrin deposit (Fig. 1A). Regions where the DFC had broken down more rapidly were filled with thrombus which did not extend into the lumen. At this time remodeling as seen by invasion of mesenchymal cells had begun, but most of the DFC matrix remained.

Around the Dacron mesh there was a persistent inflammatory response, consisting almost entirely of macrophage-monocytes and foreign body giant cells at one month. This persistent response may have limited remodeling of the matrix material (Greisler *et al*, 1986), so we developed collagen threads (Cavallaro *et al*, 1991) and knitted them into a mesh with a similar open pattern to the synthetic knit. The collagen knits were mildly crosslinked (Chvapil *et al*, 1977) so that the knit would retain strength as the remainder of matrix underwent remodeling. The collagen threads evoked a much milder reaction than Dacron as can be seen in Fig 1B where cells are growing on the threads and new capillaries can be seen in close proximity.

Much more extensive remodeling could be seen at three months. About 80% of the luminal surface was covered with flattened cells growing on a substrate of new matrix and spindle-shaped cells (Fig. 1C). Immunocytochemical staining showed that the flattened luminal cells were endothelial cells (Fig. 1D) and that many of the cells in the matrix expressed smooth muscle alpha-actin.

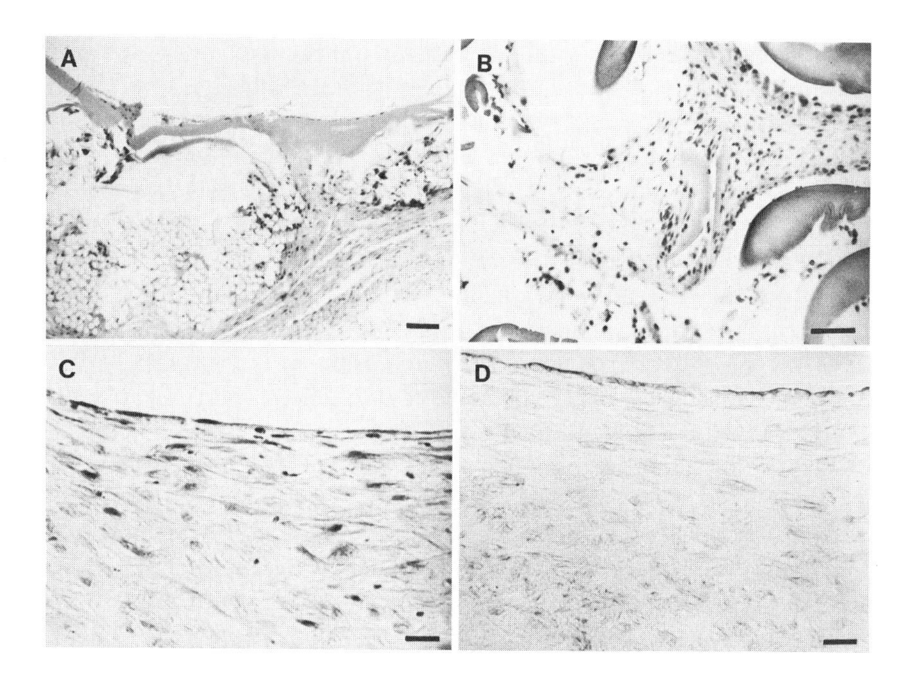

FIGURE 1. Response to DFC grafts *in vivo*. A. Lumen of DFC graft at 1 month showing minimal thrombus over intact DFC. Bar = 100 um. B. Remodeling and vascularization near collagen threads at 1 month. Bar = 50 um. C. and D. Remodeling near the lumen of DFC graft at three months. Bar = 20 um. C. Hematoxylin and eosin stain. D. Adjacent section stained for endothelial cells (*Ulex europaeus* lectin).

Laser drilling produced holes or pores in the grafts as shown in Fig. 2A. An ArF laser with a wavelength of 193nm gave very little melting around the entrance hole, while the longer wavelength KrF laser (248nm) produced a small ridge around the edge of the pore suggesting that some melting or thermal damage had occurred. *In vitro* experiments where cultured ovine endothelial cells were plated onto a laser drilled DFC matrix (with a much higher density of holes than used in the grafts) showed that the cells thrived in the holes (Fig. 2B).

Five laser-drilled DFC grafts were implanted. There was no problem with bleeding through the laser pores due to their small size. Two of these grafts were explanted for histology at month. Transmural holes, or portions thereof, were found in some sections of these grafts; such holes were never seen in the explanted DFC grafts that had not been laser-drilled. These holes were about 80um in diameter, somewhat larger than the original pores and were separated by 400-600um, as were the original laser-drilled pores (Fig. 2C). Furthermore, these holes were lined with flattened cells (Fig. 2D). Immunocytochemical staining with endothelial cell markers was suggestive but not convincing in these sections.

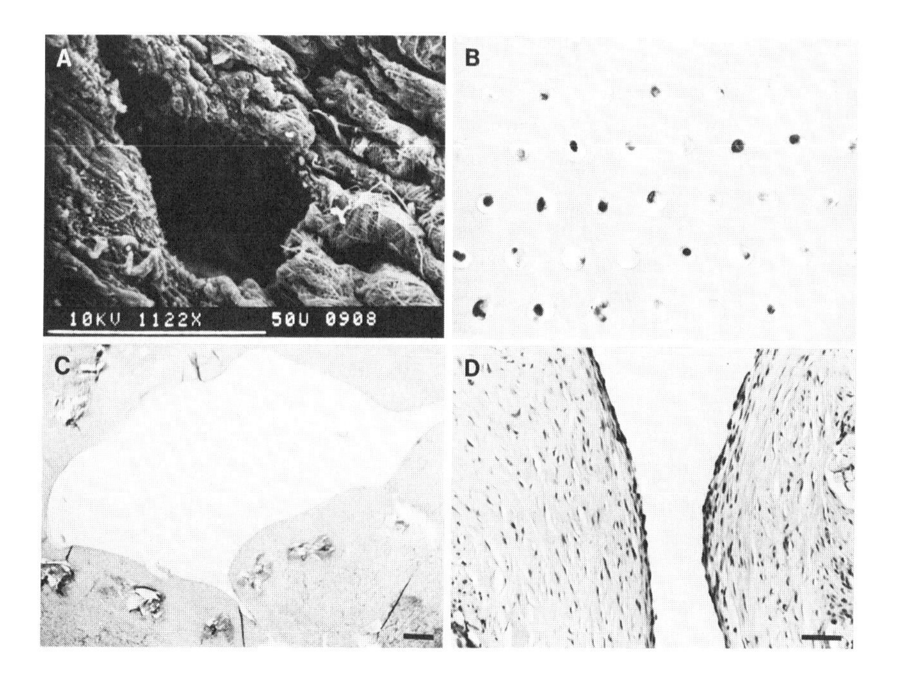

FIGURE 2. Laser drilled pores in DFC grafts. A. Scanning electron micrograph of a laser-drilled pore in a DFC matrix. Bar = 50 um. B. Cultured endothelial cells growing in laser drilled holes. C. and D. Histological sections of a laser-drilled DFC graft 1 month after implantation. Bar = 0.5 cm in C. and 50 um D.

DISCUSSION

We have made constructs from extracellular matrix molecules and have shown that these constructs have both the physical and biological properties to serve as a scaffold for guiding tissue remodeling even in a demanding setting such as a small caliber vascular graft. These initial studies illustrate the promise of a matrix engineering approach. A graft whose mode of action is to serve as a scaffold for remodeling into a tissue with cells and microcirculation provided by the host should be much more resistant to later degradation. Such long-term breakdown of materials has, to date, limited the performance of synthetic and processed biological grafts in small diameters ($\leq$ 4mm).

Overcoming the initial patency rate problem may be approached by a variety of methods including refinements in DFC fabrication, altering the knit configuration, limited crosslinking to modify remodeling rates, and addition of other components to the matrix. One step we are taking is replacing the synthetic knit with the collagen knit described above.

Type I collagen is a particularly attractive molecule for matrix engineering since it can be purified in its native in commercially useful quantities and has low immunogenicity compared to most proteins (Waksman and Mason, 1949; Canadian Hemashield Study Group, 1990). We have shown that the form in which collagen is laid down can control its hemocompatibility, that different forms (e.g. threads and sleeves) can be integrated in constructs with mechanical and handling properties superior to constructs made from either form alone, and that modifying the microarchitecture can guide the remodeling process. Thus, we feel that matrix engineering will continue to blossom into a clinically useful area of tissue engineering.

ACKNOWLEDGEMENTS

We thank Howard B. Haimes, Cynthia J. Nolte, and Joyce Bousquet for expert assistance with histology and morphology, and Valerie A. Theobald for skillful immunocytochemistry. Laser drilling was conducted by Resonetics Inc. This work was supported in part by a contract from Eli Lilly & Company.

REFERENCES

Badylak SF, Lantz GC, Coffey A and Geddes, LA (1989): Small intestine submucosa as a large diameter vascular graft in the dog. *J. Surg. Res.* 4: 74-80.

Bruns R and Gross J (1985): Transparent non-fibrilized collagen material by ultracentrifugation. *U.S. Patent 4,505,855.*

Canadian Hemashield Study Group (1990): Immunologic response to collagen-impregnated vascular grafts: A randomized prospective study. *J. Vasc. Surg.* 12: 741-746.

Carr RM, Ekstein BA, Kemp PD, O'Neil KD, Weinberg CB and Garcia JP (1991): Effects of process parameters on physical and biological properties of a small caliber vascular prosthesis. *Materials Research Society Annual Meeting.*

Cavallaro JF, Carr RM, O'Neil KD, Maresh JG and Kemp PD (1991): Effects of crosslinking on physical and biological properties of collagen threads. *Ann. Biomed. Eng.* 19: 604.

Chvapil M, Kronenthal RL and van Winkle W (1973): Medical and surgical applications of collagen. *Int. Rev. Conn. Tiss. Res.* 6; 1-61.

Chvapil M, Owen JA and Clark DS (1977): Effect of collagen crosslinking on the rate of resorption of implanted collagen tubing in rabbits. *J. Biomed. Mat. Res.* 11: 297-314.

Clowes AW, Kirkman TR and Reidy MA (1986): Mechanisms of arterial graft healing. Rapid transmural capillary ingrowth provides a source of intimal endothelium and smooth muscle in porous PTFE prostheses. *Am. J. Pathol.* 123: 220-230.

Fonkalsrud EW, Sanchez M and Zerubavel R (1978): Morphological evaluation of canine autogenous vein grafts in the arterial circulation. *Surgery* 84: 253-264.

Golden MA, Hanson SR, Kirkman TR, Schneider PA and Clowes AW (1990): Healing of ploytetrafluoroethylene arterial grafts is influenced by graft porosity. *J. Vasc. Surg.* 11: 838-845.

Greisler HP, Schwarcz TH, Ellinger J and Kim DU (1986): Dacron inhibition of arterial regenerative activities. *J. Vasc. Surg.* 3: 747-756.

Gross J and Kirk D (1958): The heat precipitation of collagen from neutral salt solutions: Some rate-regulating factors. *J. Biol. Chem.* 233: 355-360.

Jones M, Conkle DM, Ferrans VJ, Roberts WC, Levine FH, Melvin DB and Stinson EB (1973): Lesions observed in autogenous vein grafts. Light and electron microscopic evaluation. *Circulation* 47 & 48 (Suppl. III): III-198 - III-210.

Kent KC, Whittemore AD and Mannick JA (1989): Short-term and medium-term results of an all autogenous tissue policy for infrainguinal reconstruction. *J. Vasc. Surg.* 9: 107-114.

Sandusky GE, Badylak SF, Morff RJ, Johnson WD and Lantz G (1992): Histologic findings after *in vivo* placement of small intestine submucosal vascular grafts and saphenous vein grafts in the carotid artery in dogs. *Am. J. Pathol.* 140: 317-324.

Spray TL and Roberts WC (1977): Changes in saphenous veins used as aortocoronary bypass grafts. *Am. Heart J.* 94: 500-516.

Unni KK, Kottke BA, Titus JL, Frye RL, Wallace RB and Brown AL (1974): Pathologic changes in aortocoronary saphenous vein grafts. *Am. J. Cardiol.* 34: 526-532.

Waksman BH and Mason HL (1949): The antigenicity of collagen. *J. Immunol.* 63: 427-433.

Weinberg CB and Bell EB (1986): A blood vessel model constructed from collagen and cultured vascular cells. *Science* 231: 397-400.

Weinberg CB Kemp PD, O'Neil KD, Carr RM, Dennis PA, Ekstein BA, Nolte CJ, and Connolly RJ (1991): Endothelial cell and platelet interactions with smooth and fibrillar collagen surfaces. *J. Cell Biol.* 115: 442a.

Weinberg CB, Kemp PD, O'Neil KD, Carr RM, Cavallaro JF, Rosenberg M, Garcia JP (1992): Matrix engineering: Design and fabrication of a collagen vascular graft for remodeling *in vivo*. *J. Cell. Biochem.* Supplement 16F: 125.

Wesolowski SA, Fries CC, Karlson KE, DeBakey M and Sawyer PN (1961): Porosity. Primary determinant of the ultimate fate of synthetic vascular grafts. *Surgery* 50: 91-96.

BIOELASTIC MATERIALS AS MATRICES FOR TISSUE RECONSTRUCTION

Dan W. Urry
Laboratory of Molecular Biophysics
The University of Alabama at Birmingham

PHYSICAL PROPERTIES OF BIOELASTIC MATERIALS.

Bioelastic materials are elastomeric polypeptides whose origins are repeating peptide sequences in mammalian elastic fibers (Sandberg et al., 1985; Yeh et al., 1987; Indik et al., 1987). Physical properties of synthetic high polymers of the repeats have been extensively studied (Urry, 1991;1988), and they have been designed to exhibit properties not present in elastic fibers. A key property is that these elastomeric polypeptides exhibit reversible transitional behavior in which they become more-ordered on increasing the temperature through a critical temperature range (Urry, 1992). This is called an inverse temperature transition; its onset temperature is designated as T_t; and it is due to hydrophobic folding and assembly. The reverse process of hydrophobic unfolding and disassembly on lowering the temperature is well-known in proteins as cold denaturation (Privalov, 1990). The temperature, T_t, at which the transition occurs depends on the amino acid composition allowing for a T_t-based hydrophobicity scale to be developed (Urry et al., 1992a) which allows the bioelastic material to be designed to have its transition temperature set as desired within the available aqueous range or returned to a desired temperature when functional peptide sequences have been added that change the value of T_t. In spite of becoming more-ordered on raising the temperature, the more-ordered state in cross-linked matrices is a dominantly entropic elastomer (Urry, 1991) with the potential for great durability to sustain repeated stretch/relaxation cycles.

In cross-linked matrices the inverse temperature transition is seen as a thermally-driven contraction capable of performing useful mechanical work. Now any perturbation, i.e., any free energy input, that can lower the value of T_t from above to below physiological temperature becomes a means of driving contraction. This is called the ΔT_t-mechanism for free energy transduction which is summarized in Table 1 (Urry, 1993).

TABLE 1. *The ΔT_t-Mechanism of Free Energy Transduction.*

1. Bioelastic materials are elastomeric polypeptides that undergo an inverse temperature transition of hydrophobic folding and assembly on raising the temperature with the onset temperature designated as T_t.

2. In warm-blooded animals, the temperature is not raised to drive the folding, rather a number of free energy inputs can lower the transition temperature from above to below physiological temperature to drive folding and assembly. This is the ΔT_t-mechanism.

3. In cross-linked matrices the folding and assembly transition can be seen as a thermally-driven contraction. This is thermomechanical transduction by an inverse temperature transition.

4. Numerous changes in concentration of chemicals can lower the transition temperature to drive contraction. This is chemomechanical transduction by the ΔT_t-mechanism (Urry et al., 1988b; 1988a).

5. Prosthetic groups attached to the bioelastic matrix can be reduced, chemically or electrically, to lower the value of T_t and drive contraction. This is electromechanical transduction by the ΔT_t-mechanism (Urry et al., 1992b).

6. With aromatic residues present in the bioelastic material, the addition of pressure raises the transition temperature (Urry et al., 1991a) such that the release of pressure can drive contraction (Urry et al., 1993b). This is baromechanical transduction by the ΔT_t-mechanism.

7. For chemomechanical transduction the ΔT_t-mechanism is more than an order of magnitude more efficient than charge-charge interaction mechanisms (Urry, 1992).

8. These designed free energy transductions were first conceived and demonstrated using bioelastic materials. The ΔT_t-mechanism was discovered and developed on these elastomeric polypeptides.

9. A hydrophobicity scale for amino acid residues and chemical modifications thereof has been developed (Urry et al., 1992a) such that materials can be designed with any desired value of T_t.

Bioelastic materials can be designed to exhibit many free energy transductions that may be of value in tissue engineering, and they can be designed to carry out free energy transduction in response to enzymatic processes such as phosphorylation/dephosphorylation (Pattanaik et al., 1991), oxidation and reduction (Urry et al., 1992b), etc. Bioelastic materials can be designed to function much as proteins function. In fact, the ΔT_t-mechanism, discovered and demonstrated using bioelastic materials, has been proposed to be a fundamental mechanism in protein function (Urry, 1993).

BIOCOMPATIBILITY OF BIOELASTIC MATERIALS.

A set of eleven recommended biological tests for biocompatibility have been carried out and reported on (Gly-Val-Gly-Val-Pro)$_n$, i.e., Poly(GVGVP) and its 20 Mrad γ-irradiation cross-linked elastomeric matrix, i.e., X^{20}-Poly(GVGVP). This basic bioelastic material is found to be extraordinarily biocompatible (Urry et al., 1991b); it is neither mutagenic, nor toxic, nor antigenic, and in peritoneal implant studies it remains clear, optically transparent and intact with no sign of a fibrous capsule even after 6 months in the rat. It is as though this composition were totally ignored by the host.

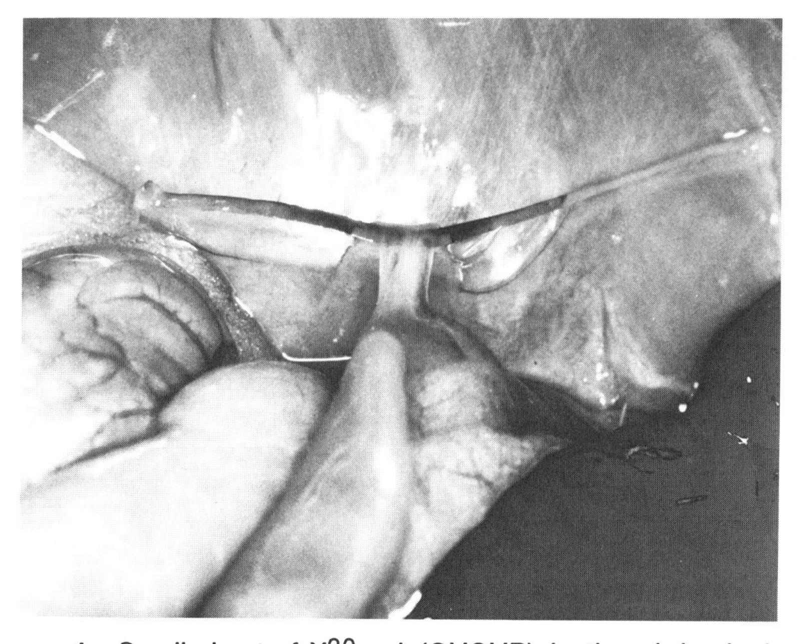

Figure 1. Small sheet of X^{20}-poly(GVGVP) in the abdominal cavity 14 days after implantation to prevent adhesion between injured wall and bowel. A small loop of adhesion has grown around, but is not adherent to, the transparent bioelastic matrix. (Reproduced with permission from Urry, et al., 1993a).

USE OF BIOELASTIC MATERIALS IN THE PREVENTION OF ADHESIONS

With these properties, the use of X^{20}-poly(GVGVP) as a barrier for the prevention of adhesion is being investigated in several animal models: in a contaminated peritoneal model in the rat (Hoban et al., 1992), in a rabbit eye model for strabismus surgery (Elsas and Urry, 1992), and, just beginning, in the calf using the total artificial heart as a bridge to transplantation in collaboration with D. Olsen at Utah.

In the contaminated peritoneal model (Hoban et al., 1992), the abdominal wall is scraped with a scalpel until it bleeds; a loop of intestine is repeatedly punctured with a hypodermic needle until it bleeds and trace contents can be extruded; and a loose suture loop is placed to hold the injured bowel in contact with the injured wall within the contaminated field. The animal is closed; the suture is cut at 7 days; and the animal is opened and examined at 14 days. In the 29 control animals, 100% of the animals exhibited adhesions. In 59 test animals, a small sheet of X^{20}-poly(GVGVP) is placed between injured wall and injured bowel and held in place by the same loose suture. For the gas sterilized bioelastic matrix, in 80% of the 29 animals, there were either no adhesions (59%) or insignificant adhesion (21%). An example of a small loop of adhesion coming around the transparent bioelastic matrix is seen in Figure 1. The basic bioelastic material is seen as a particularly innocuous matrix.

BIOELASTIC MATRICES FOR TISSUE RECONSTRUCTION

By simply adding the cell attachment sequence from fibronectin, Gly-Arg-Gly-Asp-Ser-Pro, i.e., G<u>RGD</u>SP (Pierschbacher and Ruoslahti, 1984), at the time of polymerization of the pentamers and cross-linking, the optically transparent elastomeric matrix, X^{20}-poly[20(GVGVP),(GRGDSP)], is obtained which has been transformed from a matrix without significant cell adhesion to a matrix which promotes cell attachment, cell spreading and growth to confluence (Nicol et al., 1992b). This is shown in Figure 2A and B. Interestingly, the GRGDSP sequence within the matrix functions as a vitronectin receptor rather than a fibronectin receptor (Nicol et al., 1992a).

A. X^{20}– poly(GVGVP)

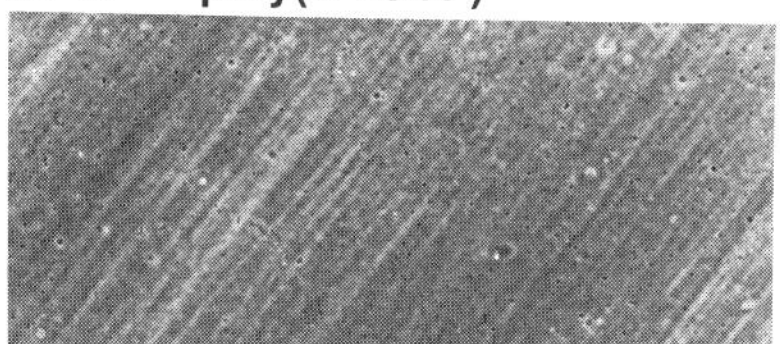

B. X^{20}– poly[40(GVGVP),(GRGDSP)]

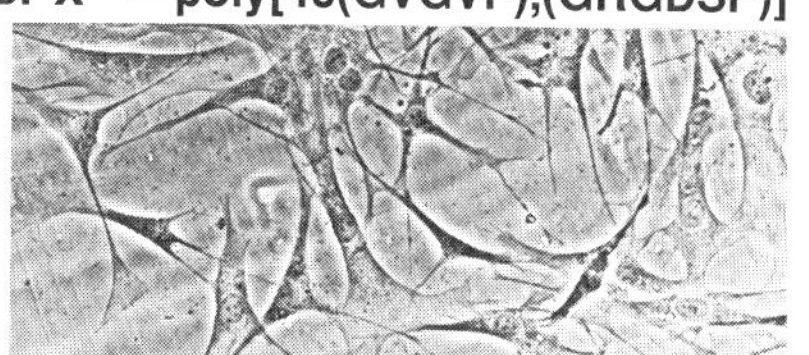

C. Unstretched Matrix

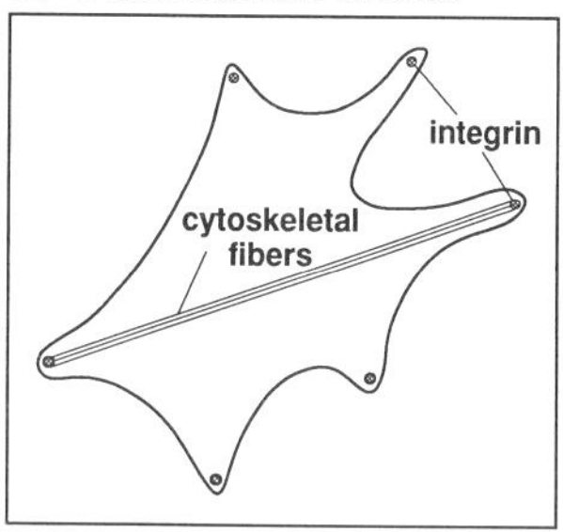

D. Stretched Matrix

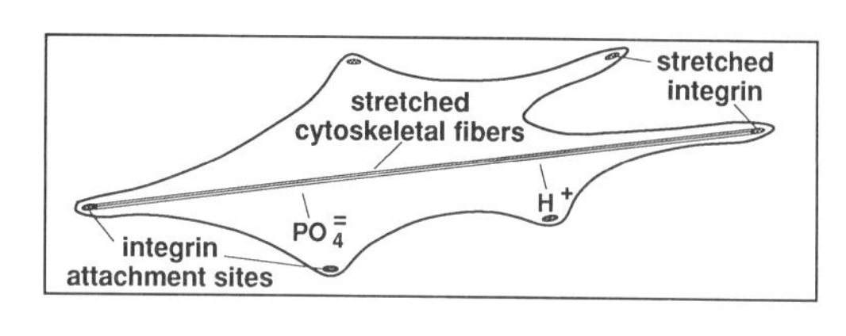

Bioelastic Matrices Containing Cell Attachment Sequences

FIGURE 2. *Bioelastic Matrices.* **A.** X^{20}-poly(GVGVP) to which cells adhere poorly or not at all. **B.** X^{20}-poly[40(GVGVP),-(GRGDSP)] to which fibroblasts and human umbilical vein and bovine aortic enndothelial cells adhere, spread and grow to confluence. **C.** Schematic of cell attached to GRGDSP through integrins. **D.** Schematic of cell on a stretched matrix showing deformed integrins and stretched cytoskeletal fibers functioning as mechanochemical transducers. **C.** and **D.** reproduced with permission from Urry 1993.

TABLE 2. *Properties of Bioelastic Materials of Relevance to Tissue Engineering.*

The basic elastomeric polypeptide and its cross-linked matrix are biocompatible and can be designed

1. to exhibit a range of elastic moduli (covering three orders of magnitude),
2. to exhibit different rates of degradation,
3. for various modes of drug and cytokine release,
4. to perform free energy transduction involving the intensive variables of mechanical force, temperature, pressure, chemical potential and electrochemical potential,
5. to contain functional enzyme sites,
6. to contain functional cell attachment sites promoting growth to confluence,

and 7. with the proper tissue elastic modulus such that the tensional forces to which the tissue is subjected can be sensed by the cell allowing for proper functioning of attached cells.

The properties of bioelastic materials which are of relevance to tissue engineering are listed in Table 2. Here, mention will only be made of item 7.

There is a growing body of evidence (see, for example, Ingber, 1991) that cells function as complex mechanochemical transducers in order to maintain a tissue appropriate for the role that it must fulfill in the body, that is, a normal cell in its natural tissue can sense the mechanical forces to which the tissue is subjected and responds chemically, for example, to turn on the genes and produce the protein required for maintaining functional tissue. That vascular smooth muscle cells attached to purified aortic elastin membranes would produce the macromolecules required for rebuilding the vascular wall only after being subjected to cyclic stretching was demonstrated by Glagov and coworkers in the 1970s (Leung et al., 1977).

If this is the case, then a central concern for tissue engineering would be to produce a synthetic matrix which would have closely the same mechanical properties as the natural tissue it replaced and which would contain sequences to which cells could attach and thereby sense the tensional forces (and their frequency) to which the prosthetic matrix is subjected. The way in which a cell could sense tensional forces is depicted in Figure 2C and D; the stretched cytoskeletal fibers or even a deformed integrin (the receptor in the cell membrane that binds at cell attachment sites) could give rise to chemical signals much as stretching the properly-designed bioelastic matrix has been shown to give rise to chemical signals. For example, stretching a Glu-containing bioelastic matrix increases the pKa of the Glu residue and results in proton uptake (Urry, 1992). Similarly, stretching could increase the free energy of a bound ATP molecule or its equivalent or a bound phosphate moiety and could lead to any number of chemical reactions.

It is the combined capacity of bioelastic materials to be designed with an elastic modulus appropriate to the tissue it is to replace and to contain cell attachment sites for the cells of that tissue that augur well for a role for bioelastic materials in tissue engineering. The perspective is to design a temporary functional scaffolding that can be remodeled by indigenous cells to become a reconstructed, properly functioning tissue.

ACKNOWLEDGMENTS

This work was supported in part by the National Institutes of Health Grant HL29578 and the Office of Naval Research Contract N00014-89-J-1970.

REFERENCES

Elsas FJ and Urry DW (1993): Synthetic polypeptide sleeve for strabismus surgery. J. Ped. Ophtha. and Strab 20: (in press).

Hoban LD, Pierce M, Quance J, Hayward I, McKee A, Gowda DC, Urry DW and Williams T (1993): The use of polypentapeptides of elastin in the prevention of postoperative adhesions. J. Surgical Res. (in press).

Indik Z, Yeh H, Ornstein-Goldstein N, Sheppard P, Anderson N, Rosenbloom J, Peltonen L and Rosenbloom J. (1987): Alternative splicing of human elastin mRNA indicated by sequence analysis of cloned genomic and complementary DNA. Proc. Natl. Acad. Sci. U.S.A. 84:5680.

Ingber D (1991): Integrins as mechanochemical transducers. Current Opinion in Cell Biology 3:841-848.

Leung DYM, Glagov S and Mathews MB (1977): A new *in vitro* system for studying cell response to mechanical stimulation. Exp. Cell Res. 109:285-298.

Nicol A, Gowda DC, Parker TM and Urry DW (1992a): Cell adhesive properties of bioelastic materials containing cell attachment sequences. In: *Biotechnol. Bioactive Polym*, Gebelein CG and Carraher, Jr., CE, eds. New York: Plenum Press.

Nicol A, Gowda DC and Urry DW (1992b): Cell adhesion and growth on synthetic elastomeric matrices containing Arg-Gly-Asp-Ser-[3]. J. Biomed. Mater. Res. 26:393-413.

Pattanaik A, Gowda DC and Urry DW (1991): Phosphorylation and dephosphorylation modulation of an inverse temperature transition. Biochem. Biophys. Res. Comm. 178:539-545.

Pierschbacher MD and Ruoslahti E. (1984): Cell attachment activity of fibronectin can be duplicated by small synthetic fragments of the molecule. Nature 309:30-33.

Privalov PL (1990): Cold denaturation of proteins. Critical Reviews in Biochemistry and Molecular Biology 25:281-305.

Sandberg L, Leslie J, Leach C, Torres V, Smith A and Smith D (1985): Elastin covalent structure as determined by solid phase amino acid sequencing. Pathol. Biol. 33: 266-274.

Urry DW (1988): Entropic elastic processes in protein mechanisms. I. Elastic structure due to an inverse temperature transition and elasticity due to internal chain dynamics. J. Protein Chem. 7:1-34.

Urry DW (1991): Thermally driven self-assembly, molecular structuring and entropic mechanisms in elastomeric polypeptides. In: *Mol. Conformation and Biol. Interactions*,

Balaram P and Ramaseshan S., eds. Indian Acad. Sci., Bangalore, India: pp. 555-583.

Urry DW (1992): Free energy transduction in polypeptides and proteins based on inverse temperature transitions. Prog. Biophys. molec. Biol. 57:23-57.

Urry DW (1993): Molecular machines: How motion and other functions of living organisms can result from reversible chemical changes. Angew. Chemie Int. Ed. Engl. 32: 819-841.

Urry DW, Harris RD and Prasad KU (1988a): Chemical potential driven contraction and relaxation by ionic strength modulation of an inverse temperature transition. J. Am. Chem. Soc. 110:3303-3305.

Urry DW, Haynes B, Zhang H, Harris RD and Prasad KU (1988b): Mechanochemical coupling in synthetic polypeptides by modulation of an inverse temperature transition. Proc. Natl. Acad. Sci. U.S.A. 85:3407-3411.

Urry DW, Hayes LC, Gowda DC and Parker TM (1991a): Pressure effect on inverse temperature transitions: Biological implications. Chem. Phys. Letters. 182:101-106.

Urry DW, Parker TM, Reid MC and Gowda DC (1991b): Biocompatibility of the bioelastic materials, poly(GVGVP) and its γ-irradiation cross-linked matrix: Summary of generic biological test results. J. Bioactive Compatible Polym. 6:263-282.

Urry DW, Gowda DC, Parker TM, Luan C-H, Reid MC, Harris CM, Pattanaik, A and Harris RD (1992a): Hydrophobicity scale for proteins based on inverse temperature transitions. Biopolymers 32:1243-1250.

Urry DW, Hayes LC, Gowda DC, Harris CM and Harris RD (1992b): Reduction-driven polypeptide folding by the ΔT_t mechanism. Biochem. Biophys. Res. Comm. 188:611-617.

Urry DW, Gowda DC, Cox BA, Hoban, LD, McKee, A and Williams T (1993a): Properties and prevention of adhesions application of bioelastic materials. Mat. Res. Soc. Symp. Proc. 292, 253-264.

Urry DW, Hayes LC, Parker TM and Harris RD (1993b): Baromechanical transduction in a model protein by the ΔT_t mechanism. Chem. Phys. Letters. 201:336-340.

Yeh H, Ornstein-Goldstein N, Indik Z, Sheppard P, Anderson N, Rosenbloom J, Cicila G, Yoon K and Rosenbloom J (1987): Sequence variation of bovine elastin mRNA due to alternative splicing. Collagen and Related Res. 7:235.

Section VII

Approaches to Allografting Engineered Cells and Tissues

INDUCTION OF IMMUNOLOGICAL UNRESPONSIVENESS IN THE ADULT ANIMAL

Richard G. Miller, Departments of Medical Biophysics and Immunology,Ontario Cancer Institute, University of Toronto

The major limitation to successful organ transplantation remains immunological rejection of the graft by the host. At present, the usual procedure for attempting to control graft rejection is to give immunosuppressive drugs (e.g. cyclosporin). However, such drugs also suppress desirable immune responses and may also have direct toxic effects. To avoid these problems, alternative strategies would be to treat the graft so that it is no longer immunogenic (treated elsewhere in this volume) or to treat the host such that it becomes specifically non-responsive against the graft. This last is the approach taken here.

The process of graft rejection is initiated by host T cells recognizing foreign determinants on the graft. Recent advances in basic immunology have greatly increased our understanding of this process and some of these advances will be briefly reviewed. The foreign determinants recognized are always associated with MHC molecules. MHC molecules are of two types, class I and class II. Class I molecules are encountered on the surfaces of almost all cells. They have a peptide-binding groove on their distal surfaces. During synthesis, class I molecules are exposed to peptides from all proteins (both foreign, e.g. viral, and self) being degraded as part of normal homeostasis in the cytoplasm of the cell. Some of these peptides can bind to the antigen-binding groove of the class I molecule and will be carried to the cell surface where they are potentially recognizable by T cells whose receptors are specific for both a particular peptide and the exterior portion of the MHC molecule forming the peptide binding groove. In a particular individual, T cells recognizing self peptides plus self MHC are eliminated during development, leaving only T cells that recognize foreign peptides plus MHC.

Class II MHC molecules have a more limited distribution, being associated primarily with cells of the immune system. They appear to have a similar structure. They also sample and bind peptides but they sample peptides from proteins that have been endocytosed and are being degraded in vacuoles inside the cell.

The MHC molecules of both classes are extremely polymorphic, particularly in the region of the antigen binding groove and it is extremely rare for two individuals to be MHC identical. When tissue is transplanted from an MHC-different donor, some host T cells appear to recognize directly the foreign donor MHC molecules. Even in the rare instance when donor and host are MHC identical, some of the self peptides presented by the MHC of the graft donor will differ from those produced by the host because of genetic variations between donor and host. These peptides will be antigenic to the host and can activate host T cells.

To become activated, a T cell must not only recognize MHC plus peptide but also receive a second non-specific signal normally produced by a specialized antigen presenting cell (APC) which delivers the non-specific signal on being recognized. These APC might be referred to as stimulatory APC. My group has been testing the hypothesis that there are other types of APC which inactivate T cells that recognize them. These cells, on being recognized, would deliver an inactivating signal. We have called such APC deletional APC or veto cells (Miller, 1980).

Veto cells can be convincingly demonstrated in tissue culture studies of the mixed lymphocyte reaction (reviewed in Miller, 1986, Fink et al.,1986, Rammensee, 1989). It is found that veto cells function through the inactivation of T cell precursors which recognize them. Resting precursor cells are not subject to being vetoed but may become sensitive at a critical stage following stimulatory activation but before differentiating into mature effector cells. Recently, it has been shown that cells carrying CD8 on their surface can act at veto cells (Sambhara and Miller, 1991). CD8 is a molecule which interacts with an invariant portion of class I MHC molecules. Thus, an activated T cell which is signalled through its antigen specific receptor (by MHC plus antigen) and also through its class I MHC molecule (by CD8) becomes permanently inactivated. There are also $CD8^-$ cells which can act as veto cells (Muraoka and Miller, 1980, Hiruma et al, 1992). Their mechanism of action is unknown.

Can veto cells be used to enhance graft survival? In principle, veto cells from the graft donor should inactivate host T cells that recognize them and thus enhance graft survival. Evidence consistent with the veto hypothesis has been discovered independently, through studies using mice and in clinical studies of kidney transplant recipients. In the mouse studies, mice were injected intravenously with foreign lymphoid cells. When these mice were subsequently tested for their ability to generate CTL in tissue culture, it was found that the anti-donor CTL response was specifically reduced (Miller and Phillips, 1976). At about the same time, it was found in clinical studies of kidney transplant recipients, that prior donor-specific blood transfusion leads to enhanced kidney graft survival (Opelz et al., 1981, van Twuyver et al., 1991). The above results can be explained by assuming the infused lymphocytes act as veto cells that inactivate T cells that recognize them.

My lab has recently been reinvestigating the question in a mouse model system. Let A and B be two MHC-different inbred mouse strains. Injection of F_1 (A x B) lymphoid cells into A inactivates both CTL precursors (Rammensee et al, 1984, Martin and Miller, 1989, Heeg and Wagner 1990) and T helper precursors (Kiziroglu and Miller, 1991) in A that can recognize B. We put a long-lasting fluorescent label on the injected cells so that they could be readily identified at later times. We found that the injected cells rapidly entered into and equilibrated with the recirculating lymphocyte pool of the host. Using a cell sorter, we found that donor-reactive host T cells were inactivated in the animal within 3 days of injection of F_1 cells as measured in tissue culture using a host cell suspension from which F_1 cells had been removed. Recovered F_1 cells, but not host cells, could transfer response reduction to a naive A recipient (Martin and Miller 1989). Many investigators have attributed the response reduction seen in this system to donor-specific suppressor cells developed by the host. The fact that F_1 cells recovered from an injected host and not host cells can transfer response reduction to a second host rules out this possibility and is consistent with the injected cells acting as veto cells that inactivate T cells that recognize them.

What is the fate of the donor-reactive T cells? Have they been killed (deleted) or rendered unresponsive (anergic)? In normal animals, it is not possible to make this distinction as the specificity of a particular T cell can only be determined by activating it and measuring its function. This problem has been overcome with the advent of transgenic mice in which it is possible to engineer a mouse in which a large fraction of the T cells carry a particular T cell receptor of known specificity. We have used mice carrying a transgenic T cell receptor (tg-TcR) reactive against male antigen (H-Y) for which there is also a monoclonal antibody that recognizes the tg-TcR (von Boehmer, 1990). This enabled us to show that injection of male cells into female mice carrying the tg-TcR led to the disappearance of the male reactive cells (Zhang et al., 1992). Male reactive cells continued to be produced in the thymus but were removed in the periphery.

Most (but not all) investigators find that an injection of F_1(AxB)lymphoid cells into A induces response reduction against B and enhances survival of a B graft. However, injection of B into A usually does not although the arguments presented above predict it should. We have recently found an explanation (Sheng-Tanner and Miller, 1992) It appears that the injected cells must enter into and remain in the host recirculating lymphocyte pool in order to induce response reduction and enhance graft survival. F_1 cells injected into A do this but cells from B injected into A do not. Instead, they are rapidly removed by host NK cells. Unlike T cells, which have receptors that recognize specific foreign determinants, NK cells appear to respond to absence of a self marker associated with (or identical to) class I MHC. Thus F_1 cells injected into A are poorly recognized by host NK cells because they carry A MHC. However, cells from strain B lack A MHC and are readily recognized. When NK cells of a recipient A mouse are in a very active state, as can be produced by infection or any other process that produces the lymphokine INNγ, even injected F_1 cells will be removed. We have directly shown that injection of F_1 into A mediates donor specific response reduction and enhances graft survival when NK cells are quiescent but not when they are active These observations have major implications for using donor lymphoid cell injection as a means of enhancing graft survival. My group is actively exploring their practical application.

REFERENCES

Fink PJ, Shimenkovitz RP and Bevan MJ (1986): Veto cells. *Ann Rev Immunol* 6: 115-147.

Heeg K and Wagner H (1990): Induction of peripheral tolerance to class I Major Histocompatibility Complex (MHC) alloantigens in adult mice: Transfused class I MHC-incompatible splenocytes veto clonal responses of antigen-reactive Lyt-2^+ T cells. *J Exp Med* 172: 719-723.

Hiruma K, Nakamura H, Henkart PA and Gress RE (1992): Clonal deletion of postthymic T cells: veto cells kill precursor cytotoxic T lymphocytes. *J Exp Med* 175: 863-868.

Kiziroglu F and Miller RG (1991): *In vivo* functional clonal deletion of recipient $CD4^+$ T helper precursor cells that can recognize class II MHC on injected donor lymphoid cells. *J Immunol* 146: 1104-1112.

Martin DR and Miller RG (1989) *In vivo* administration of histoincompatible lymphocytes leads to rapid functional deletion of cytotoxic T lymphocye precursors. *J Exp Med* 170: 679-690.

Miller RG (1980): How a specific anti-self deletion mechanism can affect the generation of the specificity repertoire. In: *The Strategies of Immune Regulation*, Sercarz EE & Cunningham AJ, eds. New York: Academic Press pp 507-512.

Miller RG (1986): The veto phenomenon and T cell regulation. *Immunol Today* 7: 112-114.

Miller RG and Phillips RA (1976): Reduction of the *in vitro* cytotoxic lymphocyte response produced by *in vivo* exposure to semiallogenic cells: Recruitment or active suppression? *J Immunol* 117: 1913-1921.

Muraoka S and Miller RG (1980): Cells in bone marrow and in T cell colonies grown from bone marrow can suppress generation of cytotoxic T lymphocytes directed against their self antigens. *J Exp Med* 152: 54-71.

Opelz G, Mickey MR and Terasaki PI (1981): Blood transfusions and kidney transplants: Remaining controversies. *Transpl Proc* 13: 136.

Rammensee H-G (1989): Veto function *in vitro* and *in vivo*. *Int Rev Immunol* 4: 177.

Rammensee H-G, Fink PJ and Bevan MJ (1984): Functional clonal deletion of class I-specific cytotoxic T lymphocytes by veto cells that express antigen. *J Immunol* 133: 2390-2396.

Sambhara SR and Miller RG (1991): Programmed cell death of T cells signalled through the T cell receptor and the α_3 domain of class I MHC. *Science* 252: 1424-1427.

Sheng-Tanner X and Miller RG (1992): Correlation between lymphocytie-induced donor-specific tolerance and donor cell recirculation. *J Exp Med* 176: 407-413.

van Twuyver E, Mooijaart RJD, ten Berge IJM, vander Horst AR, Wilmink JM, Kast WM, Melief CJM and de Waal LP (1991): Pretransplantation blood transfusion revisited. *New England J Med* 325: 1210-1213.

von Boehmer H (1990): Developmental biology of T cells in T cell receptor transgenic mice. *Ann Rev Immunol* 8: 531.

Zhang L, Martin DR, Fung-Leung W-P, Teh H-S and Miller RG (1992): Peripheral deletion of mature $CD8^+$ antigen-specific T cells after *in vivo* exposure to male antigen. *J Immunol* 148: 3740-3745.

"NEUTRAL ALLOGRAFTS" CULTURED ALLOGENEIC CELLS AS BUILDING BLOCKS OF ENGINEERED ORGANS TRANSPLANTED ACROSS MHC BARRIERS

Mireille Rosenberg
Organogenesis Inc.

To avoid the rejection of an allograft by the host, it is commonly thought that there is a need to match donor and host or to manipulate in some way either the allograft or the host immune system. Prolonged in vitro culture of organ allografts prior to transplantation may lead to a diminished rejection response (Jacobs, 1974, Lafferty et al., 1976, Lacy et al., 1990). This outcome could be due to the modification of alloantigenicity (Lafferty et al., 1976) or loss of passenger leukocytes from the graft (Lafferty et al., 1983). Neonatal tissues have been allografted without rejection, thus differentiating between antigenicity and immunogenicity in neonatal and adult tissues (Demidem et al., 1990). The immunogenic elements of an allograft have generally been thought to be cells of hematopoietic origin or small vessel endothelium rather than its parenchymal or mesenchymal cells (Dvorak et al., 1980, Faustman et al., 1984, Demidem et al., 1986). This opens the possibility of allografting cultured cells, potentially, but not necessarily of neonatal origin.

Two of the main requirements for a successful cellular allograft are the purity of the cell cultures and the maintenance of the proliferative and functional potential of the cells. Using specialized cells in a matrix of collagen, forming a scaffold that is accepted and remodeled by the body, tissues can be engineered and used as viable allografts (Bell et al., 1989). The "neutral allograft" concept relies on the fact that certain cultured cells, which do not elicit an allogeneic response, can be selected for use as building blocks in these constructs. Pure populations of cells, devoid of contaminating elements which can initiate an immune reaction, can be obtained by in vitro culture. Furthermore, these cultured cells are not recognized as foreign and are therefore not rejected.

MHC class I genes are expressed in most cells of the body. In contrast, MHC class II genes are expressed in a limited variety of cell types such as B cells, activated T cells, macrophages and dendritic cells. All of these cells are able to participate in the immune response by presenting antigen in association with class II molecules to T cells (Palacios, 1982). Other cells can be induced to express class II antigens by incubation with gamma interferon (γ-IFN) (Basham et al., 1984, Collins et al., 1984). Such induced cells may participate in antigen presentation and induce allograft rejection (Milton and Fabre, 1983, Hall et al., 1984).

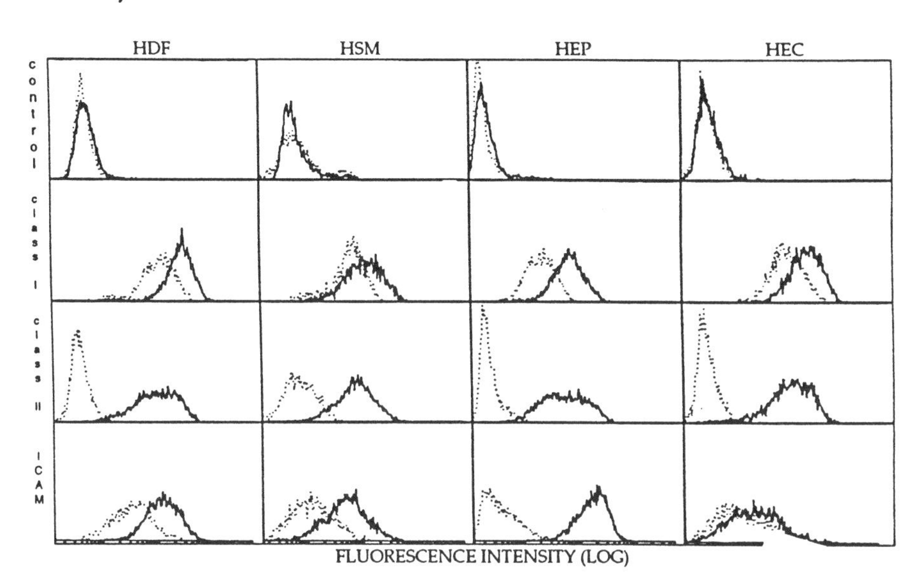

FIGURE 1.
Flow cytometry analysis of MHC antigen and ICAM-1 expression by human dermal fibroblasts (HDF), human smooth muscle cells (HSM), human epidermal cells (HEP), and human endothelial cells (HEC) after 5 days culture with (—) or without (...) 500 u/ml of γ-IFN. Staining with control mouse ascites clone NS-1, anti MHC class I clone W6/32, anti MHC class II DR clone L243, anti ICAM-1. The x axis represents the fluorescence intensity, the y axis the number of cells.

In our studies, human dermal fibroblasts, epidermal cells, smooth muscle cells, and endothelial cells were all found to express only class I antigens constitutively. γ-IFN increased this expression and induced expression of otherwise unexpressed class II antigens DR and, to a lesser extent DP and DQ. γ-IFN was also found to induce expression of ICAM-1 (Figure 1).

The mixed lymphocyte reaction (MLR) is widely considered to be a relevant in vitro model for allograft rejection and for the study of the regulatory components of such an immune response. Using the MLR as a test system, we examined the potential of such induced non-hematopoietic cells to stimulate the proliferation of unprimed T lymphocytes from peripheral blood (PBL).

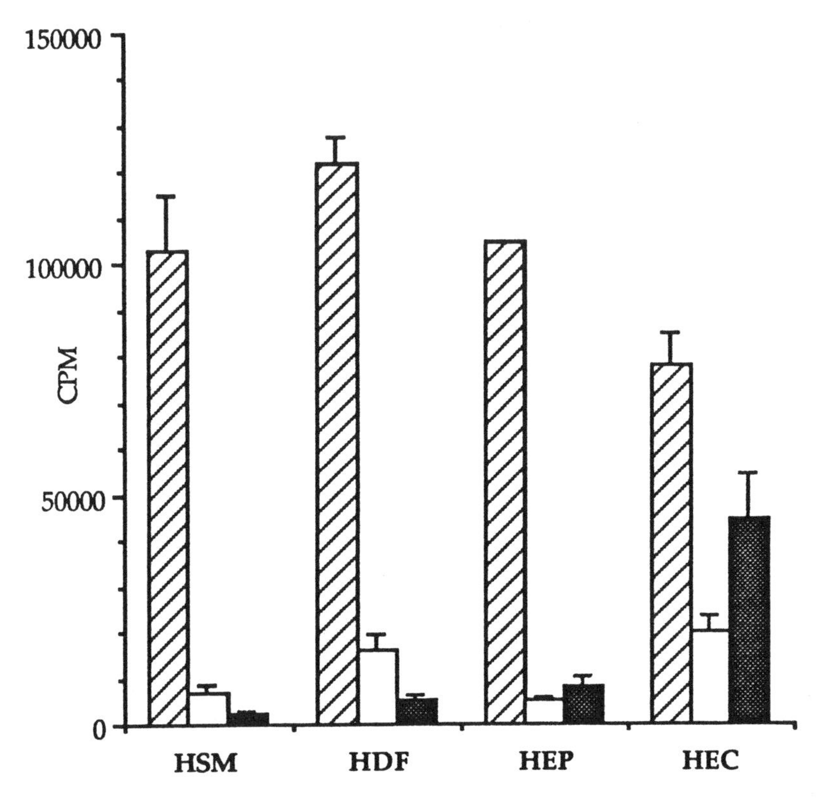

FIGURE 2.
Representative experiments showing allostimulation by human dermal fibroblasts (HDF), smooth muscle cells (HSM), epidermal cells (HEP) and endothelial cells (HEC). Stimulator cells were incubated for 5 days with (□) or without (■) 500 u/ml of γ-IFN. Stimulation by allogeneic human peripheral mononuclear cells (▨) was used as positive control. Individual cell types were tested in separate experiments. Results are presented as the means of triplicate cultures ± SD.

Endothelial cells elicited a strong proliferative alloreaction that was increased after induced expression of class II antigens on the stimulator cells. In contrast, fibroblasts, smooth muscle cells and epidermal cells failed to trigger any significant alloreaction, as measured by proliferation of T cells, regardless of class II antigen expression (Figure 2).

One must therefore distinguish between inefficient allorecognition and lack of adequate presentation. T cells recognize foreign antigens that have been processed by antigen presenting cells. MHC alloepitopes may be recognized as they are, or may require processing before their peptides can be presented to responder cells. It is not known whether the class II antigens expressed after induction on the cells studied are similar to the antigens expressed constitutively by immune cells. The MHC epitopes expressed on the induced cells were recognized by the same antibodies that recognize MHC epitopes on immune cells. Furthermore, induced class II antigens on the surface of endothelial cells were clearly immunogenic as indicated by their ability to induce proliferation.

Helpers, suppressors and cytotoxic cells are all generated during the MLR. These functional subpopulations interact via complex networks (Weaver et al., 1988). The failure of cultured fibroblasts, epidermal cells and smooth muscle cells to stimulate significant allogeneic lymphocyte proliferation may be due to the abscence or loss of accessory molecules involved in recognition. Of these, ICAM-1 is one of the most important (Marlin and Springer, 1987, Dougherty et al., 1988). Fibroblasts required transfection with both DR7 and ICAM-1 in order to stimulate freshly isolated allogeneic peripheral mononuclear cells to proliferate (Altman et al., 1989). ICAM-1 is induced in human fibroblasts, smooth muscle cells and epidermal cells by γ-IFN. However, these cells, which during the duration of the coculture coexpress MHC class II, ICAM-1 and presumably other adhesion molecules, failed to elicit allostimulation.

Cytokines such as IL-1 and IL-2 also play a role as accessory molecules in the stimulation of T cell proliferation. To investigate the possible role of soluble factors in T cell proliferation in our system, allostimulation experiments were carried out in the presence of rIL-1, r-IL-2 and r-TNF using cells preinduced by culture with γ-IFN to express MHC class II antigens. Addition of these cytokines did not influence allostimulation by human dermal fibroblasts, smooth muscle cells or human epidermal cells. IL-2 by itself stimulated a proliferative event in responder cultures which was enhanced in the presence of PBL or endothelial cells and inhibited by dermal fibroblasts, epidermal cells or smooth muscle cells. The latter cells appeared less inhibitory after incubation with gamma interferon and expression of MHC class II antigens (Figure 3 and 4). Since direct addition of IL-1, IL-2, and TNF failed to affect significantly the level of response in our cultures, it seems unlikely that any of these cytokines was the missing costimulatory factor.

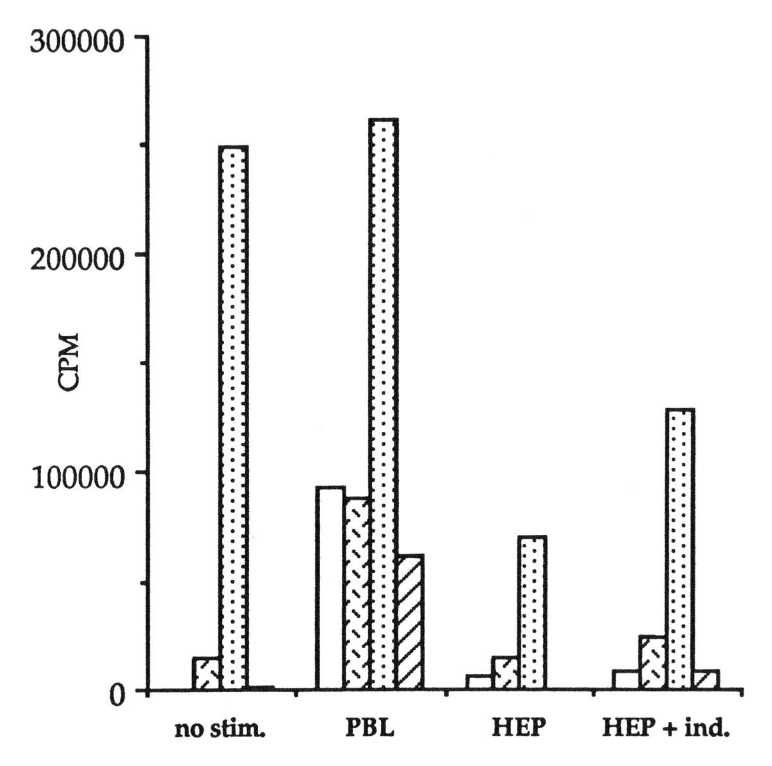

FIGURE 3.
Representative experiment showing effect of cytokines on allogeneic stimulation by cultured human epidermal cells. Control cultures were set without addition of cytokines (□); 10 u/ml of r-IL-1 (▨), IL-2 (▩) and TNF (▧) were added at the start of the MLR. Media (no stim.), peripheral blood lymphocytes (PBL), human epidermal cells (HEP) were used as stimulators after 5 days of incubation with (+ ind) or without 500 u/ml of γ-IFN. Results are presented as the means of triplicate cultures.

Allostimulation with cultured dermal fibroblasts, smooth muscle cells or epidermal cells may require additional as yet undefined accessory molecules or cytokines in order to lead to a proliferative response. Our findings confirm reports that cultured organs or cells lack immunogenicity and are in agreement with numerous studies reporting a lack of allorecognition of class II expressing parenchymal or mesenchymal cells such as fibroblasts (Umetsu et al., 1986), keratinocytes (Nickoloff et al., 1986, Gaspari et al., 1988), thyroid cells (LaRosa and Talmage, 1990), kidney nephron components (Halttunen, 1990), pancreatic cells (Markmann et al., 1988, Lo et al., 1988) and astrocytes (Sedwick et al., 1991).

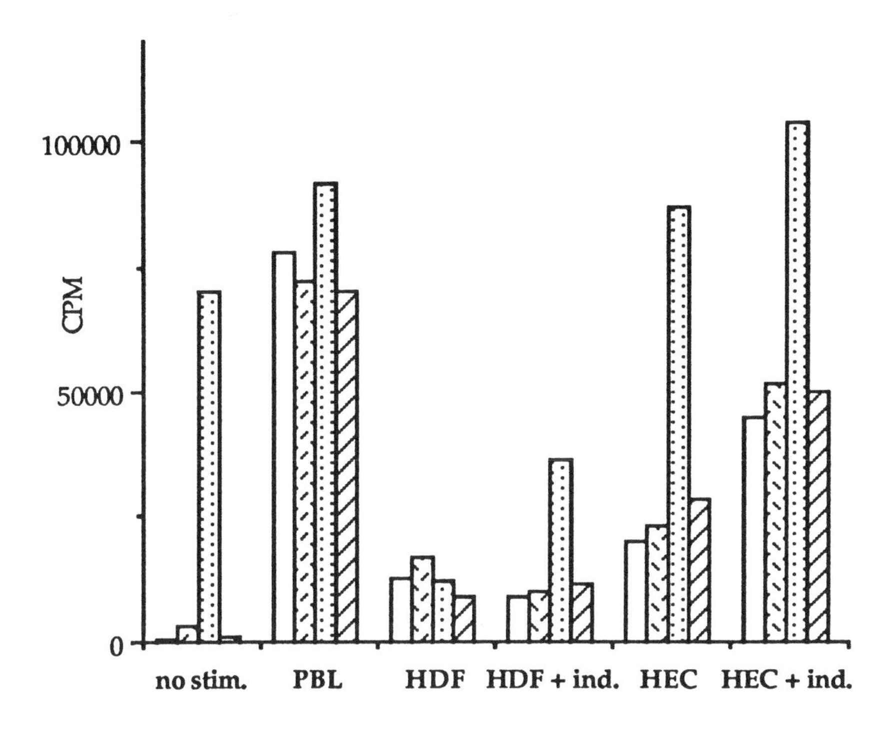

FIGURE 4.
Representative experiment showing effect of cytokines on allogeneic stimulation by cultured human cells. Control cultures were set without addition of cytokines (□); 10 u/ml of r-IL-1 (▨), IL-2 (▨) and TNF (▨) were added at the start of the MLR. Media (no stim.), peripheral blood lymphocytes (PBL), human dermal fibroblasts (HDF), and endothelial cells (HEC) were used as stimulators after 5 days of incubation with (+ ind.) or without 500 u/ml of γ-IFN. Results are presented as the means of triplicate cultures.

The experimental absence of a proliferative response, designated as a lack of allorecognition or of allostimulation, may in fact be associated with a number of different immunologically significant outcomes. The signals which fail to stimulate T cells to proliferate may nevertheless turn them on for specific cytotoxicity or the production of cytokines. Activated cerebral vascular endothelial cells expressing MHC class II antigens induced a cytotoxic response, but were unable to stimulate a proliferative response (Risau et al., 1990). A similar result has been reported with astrocytes (Sedwick et al., 1991). Although a signal may not stimulate naive cells, it may effectively turn on primed (memory) T cells. It has been reported that fibroblasts and astrocytes, even after γ-IFN induction of MHC class II expression, cannot initiate a response in naive T cells but can stimulate specifically alloreactive T cell lines (Umetsu et al., 1988). Finally, exposure of unprimed T cells to antigen presenting cells which provide improper costimulatory signals may result in a specific turn-off or clonal anergy among the responding cells (Goldsmith and Weiss, 1988, Schwartz, 1990).

The mechanism is likely to differ among the various cases listed above. In some instances simple lack of a costimulatory signal may account for failure of the proliferative response. In other cases, there may be an active suppression of proliferation of the responder cells by cell products. Although the mechanism of proliferative failure in our test system has not been identified, it is possible that the different cells examined in the present study act in different ways. However, as a pragmatic issue, the use of selected cultured allogeneic cells, which induce partial immune responses, clonal anergy or indeed fail to induce any specific response, may serve equally well in the production of engineered organ or tissue grafts which can be transplanted across major histocompatibility barriers.

REFERENCES

Altman DM, Hogg N, Trowsdale J and Wilinkson D (1989): Cotransfection of ICAM-1 and HLA-DR reconstitutes human antigen presenting cell function in mouse L cells. *Nature* 338: 512.

Basham TY, Nickoloff BJ, Merigan TC and Morhenn VB (1984): Recombinant gamma interferon induces HLA-DR expression on cultured human keratinocytes. *J Invest Dermatol* 83: 88.

Bell E, Rosenberg M, Kemp P, Parenteau NL, Haimes H, Chen J, Swiderek M, Kaplan F, Kagan D, Mason V and Boucher L (1989): Reconstitution of living organ equivalents from specialized cells and matrix biomolecules. Hybrid Artificial Organs. (Baquyard C and Dupuy B ed.) *Colloques INSERM.* 177: 13.

Collins T, Korman AJ, Wake CT, Boss JM, Kappes DJ, Fiers W, Alt KA, Gimbrone MA, Strominger JL and Pober JS (1984): Immune interferon activates multiple class II major histocompatibility complex genes and the associated invariant chain genes in human endothelial cells and dermal fibroblasts. *Proc Natl Acad Sci USA* 81: 4917.

Demidem A, Faure M, Dezutter-Dambuyant C and Thivolet J (1986): Loss of allogeneic T-cell activating ability and Langerhans cell markers in human epidermal cell cultures. *Clin Immunol Immunopathol* 38: 319.

Demidem A, Chiller JM and Kanagawa O (1990): Dissociation of antigenicity and immunogenicity of neonatal epidemal allografts in the mouse. *Transplantation* 49: 966.

Dougherty GJ, Murdoch S and Hogg N (1988): The function of human intercellular adhesion molecule (ICAM-1) in the generation of an immune response. *Eur J Immunol* 18: 35.

Dvorak HF, Mihn MC, Dvorak AM, Barnes BA and Galli SJ (1980): The microvasculature is the critical target of the immune response in vascularized skin allograft rejection. *J Invest Dermatol* 74: 280.

Faustman DL, Steinman RM, Gebel HM, Hauptfeld V, David JM and Lacy PE (1984): Prevention of rejection of murine islets allografts by pretreatement with antidendritic cell antibody. *Proc Natl Acad Sci USA* 81: 3864.

Gaspari AA, Jenkins MK and Katz SI (1988): Class II MHC bearing keratinocytes induce antigen specific unresponsiveness in hapten-specific TH1 clones. *J Immunol* 141: 2216.

Goldsmith MA and Weiss A (1988): Early signal transduction by the antigen receptor without commitement to T cell activation. *Science* 240: 1029.

Hall BM, Bishop A, Duggin GG, Horvath JS, Phillips J and Tiller DJ (1984): Increased expression of HLA-Dr antigens on renal tubular cells in renal transplants: relevance to rejection response. *Lancet* Aug 4 : 247.

Halttunen J (1990): Failure of rat kidney nephron components to induce allogeneic lymphocytes to proliferate in mixed lymphocyte kidney cell culture. *Transplantation* 50: 481.

Jacobs BB (1974): Ovarian allograft survival. Prolongation after passage in vitro. *Transplantation* 18: 454

Lacy PE, Davie JM and Fink EH (1990): The effect of islet cell culture in vitro at 24oC on graft survival and MHC antigens expression.*Transplantation* 49: 272.

Lafferty KJ, Bootes A, Dart G and Talmage DW (1976): Effect of organ culture on the survival of thyroid allografts in mice. *Transplantation* 22: 138.

Lafferty KJ, Prowse SJ, Simoneovic CJ and Warren HS (1983): Immunobiology of tissue transplantation: a return to the passenger leukocyte concept. *Ann Rev Immunol* 1: 143.

LaRosa RG and Talmage DW (1990): MHC expression on parenchymal cells of thyroid allografts is not by itself sufficient to induce rejection. *Transplantation* 49: 605.

Lo D, Burkly LC, Widera G, Cowing C, Flavell RA, Palmiter RD and Brinster RL (1988): Diabetes and tolerance induction in transgenic mice expressing class II MHC molecules in pancreatic beta cells. *Cell* 53: 159.

Markmann J, Lo D, Naji A, Palmiter RD, Brinster RL and Herber-Katz E (1988): Antigen presenting functions of class II MHC expressing pancreatic b cells. *Nature* 336: 476.

Marlin SD and Springer TA (1987): Purified intercellular adhesion molecule-1 (ICAM-1) is a ligand for lymphocyte function associated antigen-1(LFA-1). *Cell* 51: 813.

Milton AD and Fabre JW (1983): Massive induction of donor type class I and class II major histocompatibility antigens in rejection of cardiac allografts in the rat. *J Exp Med* 157: 1339.

Nickoloff BJ, Basham TY, Merigan TC, Torseth JW and Morhenn VB (1986): Human keratinocyte-lymphocyte reactions in vitro. *J Invest Dermatol* 87: 11.

Palacios R (1982): Mechanism of T cell activation: role and functional relationship of HLA-DR antigens and interleukins. *Immunol Rev* 63: 73.

Risau W, Engelhardt B and Wekerle H (1990): Immune function of the blood-brain barrier: incomplete presentation of protein (auto-) antigens by rat brain microvascular endothelium in vitro. *J Cell Biol* 110: 1757.

Schwartz RH (1990): A cell culture model for T lymphocyte clonal anergy. *Science* 248: 1349.

Sedwick JD, MoBner R, Schwender S and ter Meulen V (1991): Major histocompatilbility complex-expressing non hematopoietic astroglial cells prime only CD8+ T lymphocytes: astroglial cells as perpetuators but not initiators of CD4+ T cell responses in the central nervous system. *J Exp Med* 173: 1235.

Umetsu DT, Katzen D, Jabara HH and Geha RS (1986): Antigen presentation by human dermal fibroblasts : activation of resting T lymphocytes. *J Immunol* 136: 440.

Weaver CT, Hawrylowicz CM and Unanue ER (1988): T helper cell subsets require the expression of distinct costimulatory signals by antigen presenting cells. *Proc Natl Acad Sci USA* 85: 9699.

Index